INSTRUCTOR'S SOLUTIONS MANUAL

TO ACCOMPANY

CALCULUS
FROM GRAPHICAL, NUMERICAL, AND SYMBOLIC POINTS OF VIEW

SECOND EDITION   **VOLUME 2**

INSTRUCTOR'S SOLUTIONS MANUAL

TO ACCOMPANY

CALCULUS
FROM GRAPHICAL, NUMERICAL, AND SYMBOLIC POINTS OF VIEW

SECOND EDITION   **VOLUME 2**

OSTEBEE/ZORN

Arnold Ostebee
St. Olaf College

HOUGHTON MIFFLIN COMPANY   BOSTON NEW YORK

Sponsoring Editor: Lauren Schultz
Assistant Editor: Marika Hoe
Senior Manufacturing Coordinator: Florence Cadran
Senior Marketing Manager: Michael Busnach

Printed in the U.S.A.

ISBN: 0-618-24967-2

1 2 3 4 5 6 7 8 9 – PAT – 06 05 04 03 02

TABLE OF CONTENTS

M MULTIVARIABLE CALCULUS: A FIRST LOOK

6   *Numerical Integration*

§6.1   Approximating Integrals Numerically

1. (a) $I = \int_1^4 \frac{dx}{\sqrt{x}} = 2\sqrt{x} \Big]_1^4 = 4 - 2 = 2.$

   (b) $L_3 = \frac{4-1}{3}(f(1) + f(2) + f(3)) = \frac{1}{\sqrt{1}} + \frac{1}{\sqrt{2}} + \frac{1}{\sqrt{3}} \approx 2.28446;$

   $R_3 = \frac{4-1}{3}(f(2) + f(3) + f(4)) = \frac{1}{\sqrt{2}} + \frac{1}{\sqrt{3}} + \frac{1}{\sqrt{4}} \approx 1.78446;$

   $T_3 = (L_3 + R_3)/2 \approx 2.03446;$

   $M_3 = f\left(\frac{3}{2}\right) + f\left(\frac{5}{2}\right) + f\left(\frac{7}{2}\right) \approx 1.98347.$

   (c) $|I - L_3| \approx 0.28446;\ |I - R_3| \approx 0.21554;$

   $|I - T_3| \approx 0.03446;\ |I - M_3| \approx 0.01653.$

   (d) $S_6 = \frac{2}{3}M_3 + \frac{1}{3}T_3$

   $\approx \frac{2}{3}(1.98347) + \frac{1}{3}(2.03446)$

   $\approx 2.00047.$

   Therefore, the approximate error using S_6 is 0.00047.

2. (a) $I = \frac{x^4}{4} \Big]_0^1 = \frac{1}{4}.$

   (b) $L_4 = \left(f(0) + f(1/4) + f(1/2) + f(3/4)\right) \cdot \frac{1}{4} = \left(0^3 + (1/4)^3 + (1/2)^3 + (3/4)^3\right) \cdot \frac{1}{4} = \frac{9}{64}.$

   $R_4 = \left(f(1/4) + f(1/2) + f(3/4) + f(1)\right) \cdot \frac{1}{4} = \left((1/4)^3 + (1/2)^3 + (3/4)^3 + 1^3\right) \cdot \frac{1}{4} = \frac{25}{64}.$

   $T_4 = (L_4 + R_4)/2 = \left(f(0) + 2f(1/4) + 2f(1/2) + 2f(3/4) + f(1)\right) \cdot \frac{1}{8} = \frac{17}{64}.$

   $M_4 = \left(f(1/8) + f(3/8) + f(5/8) + f(7/8)\right) \cdot \frac{1}{4} = \frac{31}{128}.$

   (c) $|I - L_4| = \left|\frac{1}{4} - \frac{9}{64}\right| = \frac{7}{64}.$

   $|I - R_4| = \left|\frac{1}{4} - \frac{25}{64}\right| = \frac{9}{64}.$

   $|I - T_4| = \frac{1}{64}.$

   $|I - M_4| = \frac{1}{128}.$

   $|I - S_8| = |I - 2M_4/3 - T_4/3| = 0.$

3. The integrand $f(x) = \sin(x^2)$ is monotone increasing on $[0, 1]$, so all left sums underestimate and all right sums overestimate the integral.

4. (a) The integrand $f(x) = \cos(1/x)$ is monotone increasing on $[1, 3]$, so all left sums underestimate and all right sums overestimate the integral.

   (b) The integrand $f(x) = \cos(1/x)$ is concave downward on $[1, 3]$, so all trapezoid sums underestimate and all midpoint sums overestimate the integral.

5. (a) $L_{10} = 11.810$; $R_{10} = 22.530$; $T_{10} = 17.170$; $M_{10} = 16.098$; $S_{20} = 16.455$.

   (b) $|I - T_{10}| \le |L_{10} - R_{10}| = 10.72$.

   (c) $|I - S_{20}| \le |T_{10} - M_{10}| = 1.072$.

6. (a) $L_{20} \approx 0.15838$; $R_{20} \approx 0.12161$; $T_{20} \approx 0.14000$; $M_{20} \approx 0.13908$; $S_{40} \approx 0.13938$.

   (b) No, L_{20} overestimates I because the integrand is decreasing on the interval of integration.

   (c) Yes, M_{20} underestimates I because the integrand is concave up on the interval of integration.

7. (a) We'll do L_4 explicitly; the others are similar.

$$L_4 = \frac{1}{4}\big(f(0) + f(0.25) + f(0.5) + f(0.75)\big) = 1.1485.$$

   Similar calculations show:

$$R_4 = 1.1805; \quad M_2 = 1.1345; \quad T_4 = 1.1645; \quad S_4 = 1.1545.$$

   (b) Upward concavity means that T_4 must *overestimate* and M_2 must *underestimate*. Therefore I has to lie in the interval $[1.1345, 1.1645]$. (The S_4 estimate is consistent with this.)

8. (a) $L_4 = (0 + 2.6522 + 4.8755 + 6.8328) \cdot 0.75 = 10.7704$;
   $R_4 = (2.6522 + 4.8755 + 6.8328 + 8.6790) \cdot 0.75 = 17.2796$;
   $T_4 = (L_4 + R_4)/2 = 14.0250$; $M_2 = (2.6522 + 6.8328) \cdot 1.5 = 14.2275$;
   $S_4 = (2M_2 + T_2)/3 = 14.0925$.

   (b) If we assume that f is increasing and concave down on $[-1, 2]$, then all left and trapezoid sums will underestimate I, while all right and midpoints sums will overestimate I. For example, $T_4 = 14.0250 \le I \le M_2 = 14.2275$.

9. (a) Calculating I by antidifferentiation gives $I = 0.45970$ (to 5 decimals). Combining this with the tabulated data gives $I - L_{20} \approx 0.02113$; $I - R_{20} \approx -0.02094$; $I - T_{20} \approx 0.00010$; $I - M_{20} \approx -0.00005$.

   (b) L_{20} underestimates I because the integrand is increasing on $[0, 1]$.

   (c) T_{20} underestimates I because the integrand is concave down on $[0, 1]$.

10. Errors decrease in absolute value as you read down any column.

11. All entries are negative; this means that $M_n > I$ in each case.

12. (a) From technology we find $T_{100} = 1.46269$ (to 5 decimals).

    (b) Since $I = 2I_0$, a good estimate is $2 \cdot 1.46269 = 2.92539$.

13. $S_{200} = (2M_{100} + T_{100})/3 \approx 0.31027$; note that M_{100} and T_{100} are given in Example 3.

14. (a) $S_4 = \frac{1}{3}(2M_2 + T_2) = \frac{1}{6}(\cos 5 + 4\cos 5.5 + 2\cos 6 + 4\cos 6.5 + \cos 7) \approx 1.6165$.

    (b) $I = \displaystyle\int_5^7 \cos x \, dx = \sin 7 - \sin 5 \approx 1.6159$. Thus,
    $$I - S_4 \approx 1.6159 - 1.6165 \approx -0.0006;$$ note that S_4 *over*estimates I.

15. Any decreasing function will do; $f(x) = 1/x$ is one possibility.

16. Any decreasing function will do; $f(x) = 1/x$ is one possibility.

17. Any function that's concave down on $[1, 5]$ will do; $f(x) = 1 - x^2$ is one possibility.

18. Any function that's concave down on $[1, 5]$ will do; $f(x) = 1 - x^2$ is one possibility.

19. (a) Upward concavity means that the graph "sags beneath" the straight line segments that define the trapezoid rule.

    (b) Yes: T_n still overestimates I for all $n \geq 1$. Concavity matters, but not positivity or negativity of f.

20. (a) Since f is increasing over the interval $[-2, 5]$, $L_{10} \leq I \leq R_{10}$. Therefore,
    $$L_{10} - L_{10} \leq I - L_{10} \leq R_{10} - L_{10} \implies |I - L_{10}| \leq R_{10} - L_{10}.$$

    (b) $T_{10} = (L_{10} + R_{10})/2 = 9.495$.

    (c) T_{10} is the midpoint of the interval $[L_{10}, R_{10}]$, which has "radius" 0.0818. Since I lies in this interval, it can't be farther than 0.0818 from the midpoint of the interval.

21. By definition, $L_n = \big(f(x_0) + f(x_1) + \cdots + f(x_{n-1})\big)\Delta x$ and
    $R_n = \big(f(x_1) + f(x_2) + \cdots + f(x_n)\big)\Delta x$. Adding these formulas and dividing by two gives
    $$T_n = \left(\frac{f(x_0)}{2} + f(x_1) + \cdots + f(x_{n-1}) + \frac{f(x_n)}{2}\right)\Delta x,$$ as desired.

22. For *any* function, T_n is the midpoint of the interval with endpoints L_n and R_n. This interval has "radius" $|L_n - R_n|/2$. If f is monotone then we know that I lies in this same interval, and so it can't lie farther than $|L_n - R_n|/2$ (the radius) away from the midpoint T_n.

23. For *any* function, S_{2n} lies between M_n and T_n. This interval has length $|M_n - T_n|$. If f never changes direction of concavity, then I must lie in this same interval. Thus both S_{2n} and I lie in the same interval, and so can't differ by more than $|M_n - T_n|$, the interval's length.

24. For *any* function, T_n lies between L_n and R_n. For a *decreasing* function, we know $R_n \leq L_n$, and so $R_n \leq T_n \leq L_n$.

25. A picture similar to that in the discussion of Theorem 2 makes this clear.

26. Because the function is increasing and concave down, L_{30} and T_{30} underestimate I; M_{30} and R_{30} overestimate I. Thus we have $L_{30} < T_{30} < I < M_{30} < R_{30}$.

27. Let $I = \displaystyle\int_1^7 f(x)\,dx$. Since f is positive on the interval of integration, all five estimates are positive numbers. Since f is increasing on the interval of integration, $L_{100} \leq I \leq R_{100}$. Since f is concave down on the interval of integration, $T_{100} \leq I \leq M_{100}$. Furthermore, since T_{100} is the average of L_{100} and R_{100}, $L_{100} \leq T_{100} \leq R_{100}$. Also, S_{200} is a weighted average of T_{100} and M_{100}, so $T_{100} \leq S_{200} \leq M_{100}$. Therefore,

$$L_{100} \leq T_{100} \leq S_{200} \leq M_{100} \leq R_{100}.$$

28. Let $I = \int_a^b f(x)\,dx$. Since f is increasing on $[a, b]$, the left rule *underestimates* I, the right rule *overestimates*, and the other two lie in between. Since f is concave up on the interval $[a, b]$, $M_n \leq T_n$ must be true. Thus, $L_n \leq M_n \leq T_n \leq R_n$ so $L_n = 8.52974$, $M_n = 9.71090$, $T_n = 9.74890$, and $R_n = 11.04407$.

29. Must: Left sums underestimate.

30. May: It depends on the direction of concavity.

31. Must: On each subinterval f is larger at the right endpoint than at the left endpoint.

32. Must: On each subinterval f is larger at the midpoint than at the left endpoint.

33. Must: We know $L_n < R_n$, and T_n is the average of L_n and R_n.

34. May: It depends on the direction of concavity.

35. May: It depends on whether f is increasing or decreasing.

36. Must: Upward concavity means that midpoint sums underestimate.

37. May: It depends on whether f is increasing or decreasing.

38. May: It depends on whether f is increasing or decreasing.

39. May: It depends on whether f is increasing or decreasing.

40. Cannot: Upward concavity means that midpoint sums underestimate and trapezoid sums overestimate.

41. Must: f decreasing means that all right sums underestimate.

42. Must: f concave down means that all trapezoid sums underestimate.

43. Cannot: f decreasing means that left sums overestimate and right sums underestimate.

44. Cannot: We know $L_n > R_n$, and T_n is the average of L_n and R_n.

45. Must: On each subinterval f is larger at the midpoint than at the right endpoint.

46. Must: Downward concavity means that trapezoid sums underestimate and midpoint sums overestimate.

47. Must: Because $f' > 0$, we know f is increasing, so left sums underestimate.

48. **Cannot** be true. Since f is concave upwards on the interval of integration, any trapezoid rule estimate overestimates I (i.e., $I - T_n < 0$).

49. **Must** be true. Since f is increasing on the interval of integration $M_n < R_n$ for every n. [If f is increasing on an interval $[a, b]$, then $m = (a + b)/2$ is the midpoint of the interval and $f(a) < f(m) < f(b)$.]

50. **Cannot** be true. Because f is concave upwards on the interval of integration, $0 \le M_{10} \le I \le T_{10}$. Therefore, $M_{10} \le S_{20} \le T_{10}$.

51. The integrand's concavity changes twice on the interval $[0, 2]$, once near $x = 0.808$ and again near $x = 1.814$. Breaking up the integral at these points allows the sort of trapping desired.

52. (a) The integrand is increasing until $x = 1$ and decreasing thereafter, so the integrals $\int_0^1 xe^{-x}\, dx$ and $\int_1^4 xe^{-x}\, dx$ can be trapped separately by left and right sums.

    (b) The integrand is concave down until $x = 2$ and concave up thereafter, so the integrals $\int_0^2 xe^{-x}\, dx$ and $\int_2^4 xe^{-x}\, dx$ can be trapped separately by trapezoid and midpoint sums.

53. (a) Since f is linear, the trapezoid approximation "fits" f exactly over each subinterval.

    (b) A picture shows that the trapezoidal area is equal to a rectangular area.

    (c) The preceding parts show that, for a linear function, the midpoint and trapezoid estimates are identical.

54. The key idea is that for equally-spaced points $a = x_0 < x_1 < x_2 < \cdots < x_n = b$,

$$L_n = f(x_0)\Delta x + f(x_1)\Delta x + f(x_2)\Delta x + \cdots + f(x_{n-1})\Delta x,$$

while

$$R_n = f(x_1)\Delta x + f(x_2)\Delta x + f(x_3)\Delta x + \cdots + f(x_n)\Delta x,$$

where $\Delta x = (b - a)/n$. Thus, by algebra,

$$R_n - L_n = f(x_n)\Delta x - f(x_0)\Delta x = \big(f(b) - f(a)\big)\Delta x = \big(f(b) - f(a)\big)\frac{b - a}{n}$$

or, equivalently, $R_n = L_n + \big[f(b) - f(a)\big] \cdot \dfrac{(b - a)}{n}$.

55. Recall that $T_n = (L_n + R_n)/2$. Into this, substitute the expression for R_n derived in the previous exercise.

56. Suppose first that $n = 1$. Because f is increasing, the graph lies inside the rectangle with lower left corner $(a, f(a))$ and upper right corner $(b, f(b))$. (Draw a picture!) Because f is concave up, the f-graph sags below the diagonal line from $(a, f(a))$; this implies that L_1 estimates I better than R_1. The same argument (applied to each subinterval) works if $n > 1$.

57. Suppose first that $n = 1$. Because f is increasing, the f-graph lies inside the rectangle with lower left corner $(a, f(a))$ and upper right corner $(b, f(b))$. (Draw a picture!) Because f is concave up, the f-graph sags below the diagonal line from $(a, f(a))$. A close look at the picture shows that the midpoint estimate M_1 approximates I better than both L_1 and R_1. The same argument (applied to each subinterval) works if $n > 1$.

58. The information given implies that f is increasing and concave down on the interval $[a, b]$. A sketch shows that the approximation error made by L_n includes all of the area corresponding to the approximation error made by T_n and more.

    An algebraic proof of this result is also possible. Since f is (strictly) increasing and (strictly) concave down on the interval of integration, $L_n < I < R_n$ and $I - T_n > 0$. Thus,

    $$(I - L_n) - (I - T_n) = T_n - L_n = \tfrac{1}{2}(R_n - L_n) > 0$$

    which implies that $|I - T_n| < |I - L_n|$.

59. The information given implies that f is increasing and concave up on the interval $[a, b]$. A sketch shows that the approximation error made by R_n includes all of the area corresponding to the approximation error made by T_n and more. Thus the answer is "no."

60. Since $F''(x) = f'(x) \geq 0$ on $[a, b]$, T_{100} overestimates the value of I.

61. Since f' is negative on $[a, b]$, f is decreasing; therefore, $R_n \leq I \leq L_n$, $M_n \leq L_n$, and $R_n \leq T_n$. Since f' is decreasing on $[a, b]$, f is concave down; therefore, $T_n \leq I \leq M_n$. Putting these pieces together, we have $R_n \leq T_n \leq I \leq M_n \leq L_n$.

62. Since $f(x)$ is decreasing, the integrand is increasing, so L_n must underestimate I.

63. The picture shows that $(f'(x))^2$ is increasing; thus, so is the integrand. This implies that R_n overestimates the integral.

64. Note that $e^{-f(x)}$ has second derivative $(f'(x))^2 e^{-f(x)} - e^{-f(x)} f''(x)$. Since $e^{-f(x)} > 0$ for all x and the graphs show $f'(x) < 0$ and $f''(x) < 0$, it follows that the integrand's second derivative is positive. Thus, M_n underestimates I.

65. The graph shows $f'(x)$; since $f' < 0$ and f' is increasing, the function f itself must be decreasing and concave upward. It follows that $R_n \leq M_n \leq S_{2n} \leq T_n \leq L_n$. (Note that S_{2n} lies between M_n and T_n for every integral.)

66. Let $g(x) = xf(x)$ be the integrand; note $g'(x) = f(x) + xf'(x)$. The graph shows $f'(x) < 0$ for $0 \leq x \leq 4$. Thus f is decreasing; since $f(0) = 0$ we have $f(x) < 0$ for $x > 0$. Putting this information together shows that $g'(x) = f(x) + xf'(x) < 0$ for $x \geq 0$. Thus g is decreasing, so we have $R_n < I < L_n$.

67. Let $g(x) = (f(x))^2$; then $g'(x) = 2f(x)f'(x)$. Since $f(x) < 0$ and $f'(x) < 0$ for $x > 0$, we have $g'(x) > 0$, so g is increasing. Thus, $L_n \leq I \leq R_n$.

68. Let $g(x) = e^{f(x)}$; then $g''(x) = e^{f(x)}(f'(x)^2 + f''(x))$. The graph shows $f''(x) > 0$, so $g''(x) > 0$ on [0, 4]. Thus, $M_n \leq I \leq T_n$ for all n.

69. The graph shows $f'' > 0$, so f is concave up. Thus M_n must underestimate I.

70. The inequality $I \leq L_{200}$ holds if f is increasing, but not if f is decreasing. Either is possible.

71. The graph shows that f'' is positive—hence f is *concave up*—over the interval of integration. $L_{100} \leq R_{100}$ could be true or false, depending on whether f is increasing or decreasing. The graph of f'' doesn't tell which is the case.

72. $T_{200} \leq M_{200}$ is false: If the graph of f is concave up, T_n must *overestimate* I and M_n must *underestimate*. Thus $T_n > M_n$ for any n.

73. $M_{50} \leq L_{50}$ could be true or false, depending on whether f is increasing or decreasing. The graph of f'' doesn't tell which is the case.

74. Since S_{2n} always lies between M_n and T_n, we need only decide which of M_n and T_n is larger. Because f is concave up (see the graph of f''), we see $M_n \leq T_n$. Thus $M_{100} \leq S_{200} \leq T_{100}$.

§6.2 Error Bounds for Approximating Sums

1. (a) Plotting f'' over [0, 3] shows that $K_2 = 32$ is suitable. Plugging $K_2 = 32$, $a = 0$, $b = 3$, and $n = 50$ into the error bound formula gives $|I - M_{50}| \le 0.0144$, which is well below 0.02.

    (b) Here $M_{50} \approx 0.774$; part (a) implies that I lies within 0.02 of this number. That is, $0.754 \le I \le 0.794$.

2. An error bound says that some quantity can be in error by no *more* than a given amount. Error bounds are "pessimistic" in that they represent worst-case outcomes.

3. All three integrands are linear functions, so their derivatives are constants. The values of $|f'|$ are, respectively, 1, 2, and 4; the absolute errors committed by L_4 (0.125, 0.250, and 0.500) are in the same proportion.

4. Note that (a) and (b) are both clear graphically. The following arguments use Theorem 3.

    (a) For a linear function $f'' = 0$, so Theorem 3 gives $|T_n - I| = 0$ for all n.

    (b) For a linear function $f'' = 0$, so Theorem 3 gives $|M_n - I| = 0$.

5. (a) Using $K_1 = 5$, $a = 0$, $b = 2$, and $n = 8$ in Theorem 3 gives $|I - R_8| \le 5/4$.

    (b) Using $K_2 = 0$ in Theorem 3 gives $|I - T_n| = 0$ for all n.

6. (a) Using $K_1 = 24$, $a = 0$, $b = 4$, and $n = 5$ in Theorem 3 gives $|I - L_5| \le 38.4$.

    (b) Using $K_2 = 6$, $a = 0$, $b = 4$, and $n = 5$ in Theorem 3 gives $|I - M_5| \le 0.64$.

7. (a) Plotting f' shows $|f'(x)| \le 1.3$ on [0, 1]; thus any $K_1 \ge 1.3$ is acceptable.

    (b) Plotting f'' shows $|f''(x)| > 2$ near $x = 1$, so 2 is not an acceptable value for K_2.

8. The condition $-4 \le f'(x) \le 3$ if $1 \le x \le 2$ implies that K_1 must be a number greater than $|-4| = 4$. Thus, Theorem 3 guarantees that $|I - L_n| \le 4/(2n)$.

9. (a) $I = \int_1^4 \dfrac{dx}{\sqrt{x}} = 2\sqrt{x}\,\Big]_1^4 = 4 - 2 = 2.$

    (b) Calculation gives $L_3 \approx 2.28446$ and $R_3 \approx 1.78446$. Comparing these with the exact answer, 2, gives $|I - L_3| \approx 0.28446$ and $|I - R_3| \approx 0.21554$. The error bound from Theorem 3 is found by using $K_1 = 0.5$, $a = 1$, $b = 4$, and $n = 3$; the result is 0.75, which is well above the actual errors committed.

    (c) Calculation gives $T_3 \approx 2.03446$ and $M_3 \approx 1.98347$. Comparing these with the exact answer, 2, gives approximation errors $|I - T_3| \approx 0.03446$ and $|I - M_3| \approx 0.01653$. The error bounds from Theorem 3 are found by using $K_2 = 0.75$, $a = 1$, $b = 4$, and $n = 3$. This gives $|I - T_3| \approx 0.18750$ and $|I - M_3| \approx 0.09375$, which are well above the actual errors committed.

(d) With $K_2 = 0.75$, $a = 1$, and $b = 4$, Theorem 3 says that $|I - M_n| \leq \dfrac{0.84375}{n^2}$. The last quantity is less than 0.005 if $n \geq 13$; this guarantees that M_n has two decimal place accuracy if $n \geq 13$.

10. (a) $I = \left. \dfrac{x^4}{4} \right]_0^1 = \dfrac{1}{4}.$

    (b) Calculation gives $L_4 = 9/64$ and $R_4 = 25/64$. The exact integral is 16/64, so the actual absolute errors committed are 7/64 and 9/64, respectively. To find the error bounds for L_4 and R_4 from Theorem 3 we can plug in $K_1 = 3$, $a = 0$, $b = 1$, and $n = 4$. The result is 3/8, well above the actual errors committed.

    (c) Calculation gives $T_4 \approx 0.26563$ and $M_4 \approx 0.24219$. The exact integral is 0.25, so the actual absolute errors committed are about 0.01563 for T_4 and 0.00781 for M_4. To find the error bounds for T_4 and M_4 from Theorem 3 we can plug in $K_2 = 6$, $a = 0$, $b = 1$, and $n = 4$, to get error bounds of $1/32 \approx 0.032$ for T_4 and $1/64 \approx 0.016$ for M_4. Both bounds are well above the actual errors committed.

    (d) With $K_2 = 6$, $a = 0$, and $b = 1$, Theorem 3 gives $|I - M_n| \leq \dfrac{1}{4n^2}$. The last quantity is less than 0.005 if $n \geq 8$; this guarantees that M_n has two decimal place accuracy if $n \geq 8$.

11. For $\int_0^3 e^{-x^2}\, dx$ we have $f'(x) = -\exp(-x^2)2x$. On [0, 3], $|f'(x)| \leq 0.8578$, so $K_1 = 0.8578$ works. Hence we need $n \geq 0.8578 \cdot 3^2 \cdot 100 = 772.02$ so $n \geq 772$. (If K_1 is approximated, slightly different results occur.)

12. For $\int_0^2 \sin\left(x^2\right)\, dx$, $K_1 = 4$ works, so any $n \geq 4 \cdot 2^2 \cdot 100 = 1600$ will do.

13. For $\int_0^1 \left(1 + x^2\right)^{-1}\, dx$, $K_1 = 0.65$ works so any $n \geq 0.65 \cdot 1^2 \cdot 100 = 65$ will do.

14. For $\int_1^{10} \sin(1/x)\, dx$, $K_1 = \cos 1$ works so any $n \geq \cos 1 \cdot 9^2 \cdot 100 \approx 4376.45$ will do.

15. For $\int_0^3 e^{-x^2}\, dx$, $K_2 = 2$ works. Setting $K_2 = 2$, $a = 0$, and $b = 3$ in Theorem 3 shows that we want n so $\dfrac{2 \cdot 3^3}{24n^2} < 0.005$. Solving for n gives $n \geq 22$.

16. For $\int_0^2 \sin(x^2)\, dx$, $K_2 = 11$ works. Setting $K_2 = 11$, $a = 0$, and $b = 2$ in Theorem 3 shows that we want n so $\dfrac{11 \cdot 2^3}{24n^2} < 0.005$. Solving for n gives $n \geq 28$.

17. For $\int_0^1 (1 + x^2)^{-1}\, dx$, $K_2 = 2$ works. Setting $K_2 = 2$, $a = 0$, and $b = 1$ in Theorem 3 shows that we want n so $\dfrac{2 \cdot 1^3}{24n^2} < 0.005$. Solving for n gives $n \geq 5$.

18. For $\int_1^{10} \sin(1/x)\, dx$, $K_2 = 0.45$ works. Setting $K_2 = 0.45$, $a = 1$, and $b = 10$ in Theorem 3 shows that we want n so $\dfrac{0.45 \cdot 9^3}{24n^2} < 0.005$. Solving for n gives $n \geq 53$.

19. Any linear and increasing function will work; $f(x) = x$ is one possibility.

20. Any linear and decreasing function will work; $f(x) = -x$ is one possibility.

21. If $f(x) = x^2$, then f is concave up, so M_n underestimates I and the approximation error is as bad as Theorem 3 allows.

22. If $f(x) = -x^2$, then f is concave down, so M_n overestimates I and the approximation error is as bad as Theorem 3 allows.

23. (a) Doubling n halves the error bound.

    (b) For M_n and M_{2n}, doubling n reduces the error bound by a factor of 4.

    (c) Parts (a) and (b) suggest that, as a rule, the midpoint and trapezoid rules perform better than the left and right rules.

24. (a) Multiplying n by 10 in the error bound for L_n increases the denominator by a factor of 10, so we get one more decimal place of guaranteed accuracy.

    (b) Multiplying n by 10 in the error bound for M_n increases the denominator by a factor of 100, so we get two more extra decimal places of accuracy.

    (c) Parts (a) and (b) suggest that for most purposes M_n is a better method than L_n.

25. With no bound given for K_1 we have no accuracy guarantees.

26. With no bound given or available for K_2 we have no accuracy guarantees.

27. (a) For $\int_1^{11} \cos(1/x)\,dx$, $K_1 = 0.8415$ works. Setting $K_1 = 0.8415$, $a = 1$, and $b = 11$ in Theorem 3 shows that we want n so $\dfrac{0.8415 \cdot 10^2}{2n} < 0.005$. Solving for n gives $n \geq 8415$.

    (b) For $\int_1^6 \cos(1/x)\,dx$, $K_1 = 0.8415$ works. Setting $K_1 = 0.8415$, $a = 1$, and $b = 6$ in Theorem 3 shows that we want n so $\dfrac{0.8415 \cdot 5^2}{2n} < 0.004$. Solving for n gives $n \geq 2630$.

    (c) For $\int_6^{11} \cos(1/x)\,dx$, $K_1 = 0.005$ works. Setting $K_1 = 0.005$, $a = 6$, and $b = 11$ in Theorem 3 shows that we want n so $\dfrac{0.005 \cdot 5^2}{2n} < 0.001$. Solving for n gives $n \geq 63$.

    (d) Adding estimates for the two partial integrals gives an estimate for the total integral; the total error is guaranteed less than $0.004 + 0.001 = 0.005$.

    (e) Using the results above gives a good estimate with about 2700 values of f—a lot less than the original 8415 values of f.

28. (a) For $\int_0^5 e^{-x^2}\,dx$, $K_1 = 0.86$ works. Setting $K_1 = 0.86$, $a = 0$, and $b = 5$ in Theorem 3 shows that we want n so $\dfrac{0.86 \cdot 5^2}{2n} < 0.01$. Solving for n gives $n = 1075$.

(b) For $\int_0^2 e^{-x^2}\,dx$, $K_1 = 0.86$ works. Setting $K_1 = 0.86$, $a = 0$, and $b = 2$ in Theorem 3 shows that we want n so $\dfrac{0.86 \cdot 2^2}{2n} < 0.005$. Solving for n gives $n = 344$.

(c) For $\int_2^5 e^{-x^2}\,dx$, $K_1 = 0.074$ works. Setting $K_1 = 0.074$, $a = 2$, and $b = 5$ in Theorem 3 shows we want n so $\dfrac{0.074 \cdot 3^2}{2n} < 0.005$. Solving for n gives $n = 67$.

(d) The sum of the estimates from parts (b) and (c) approximates I within $0.005 + 0.005 = 0.01$.

(e) The estimate computed in part (d) requires less than half as much computation effort as the estimate in part (a) — 411 versus 1075 function evaluations.

29. (a) $I = \pi$ because the integral gives the area of the "northeast quadrant" of a circle of radius 2.

(b) $L_{10} \approx 3.3045$; $|I - L_{10}| \approx 0.1629$.

(c) Theorem 3 doesn't give a good bound here because we can't compute K_1 — $f'(x)$ is unbounded on the interval $(0, 2)$.

30. The graph shows that $\left|f'(x)\right| \le 9$ if $0 \le x \le 4$, so the error bound inequality

$$|I - L_n| \le \frac{9 \cdot 4^2}{2n} \le 0.0001$$

implies that any $n \ge 720{,}000$ works.

31. Since $F'' = f'$, the graph shows that we can take $K_2 = 9$. Setting $K_2 = 9$, $a = 1$, and $b = 2$ in Theorem 3 shows we want n so $\dfrac{9 \cdot 1^3}{24n^2} < 0.01$. Solving for n gives $n \ge 7$.

32. Since $F' = f$, we need to bound $|f(x)|$ for $2 \le x \le 4$. The graph shows $|f'(x)| \le 4$ on $2 \le x \le 4$, so $f(2) = 0$ and the speed limit law imply that $|f(x)| \le 8$ over the same interval. Thus, $\left|L_n - \int_2^4 F(x)\,dx\right| \le \frac{8 \cdot 2^2}{2n} \le 0.01$ if $n \ge 1600$.

33. Since $\left|f''(x)\right| \le 2.5$ if $0 \le x \le 5$, $\left|T_n - \int_a^b f(x)\,dx\right| \le 2.5 \cdot 5^3/12n^2$. The expression on the right is less than 0.001 if $n \ge 162$.

34. (a) $R_4 = 0.75 \cdot (26.522 + 48.755 + 68.328 + 86.790) = 172.79625$.
$|I - R_4| \le 0.75 \cdot |86.790 - 2.0000| = 63.5925$.

(b) $T_4 = \frac{1}{2}(L_4 + R_4) = \frac{1}{2} \cdot 0.75 \cdot (2 - 86.790) + R_4 = 141.00$.
From the graph, it is apparent that $\left|f''(x)\right| < 6$ if $-1 \le x \le 2$ so, taking $K_2 = 6$, we have

$$|I - T_4| \le \frac{6 \cdot 3^3}{12 \cdot 4^2} = 0.84375.$$

35. Cannot: To estimate the error committed by L_{100}, we use $K_1 = 5$, so
    $|I - L_{100}| \leq 5 \cdot 10^2/200 = 2.5$. Since $I = 9$, we must have $L_{100} < 11.5$.

36. Cannot: To estimate the error committed by M_{100}, we use $K_2 = 4$, so
    $|I - M_{100}| \leq 4 \cdot 10^3/(24 \cdot 100^2) = 1/60 \approx 0.017$. Since $I = 9$, we must have
    $M_{100} > 9 - 0.017 = 8.983$.

37. Cannot: Because h is concave up, T_n must overestimate $I = 9$.

38. Must: Because h is concave up, T_n must overestimate $I = 9$, so $I - T_{100} < 0$.

39. For a monotone function, I lies *between* L_n and R_n. The distance between L_n and R_n is
    $|R_n - L_n|$, and a little algebra shows that this difference is $|f(b) - f(a)| \cdot (b - a)/n$.

40. The idea is that for any monotone function, both M_n and I must lie somewhere *between* L_n
    and R_n. The distance between L_n and R_n is $|R_n - L_n|$, and a little algebra shows that this
    difference is $|f(b) - f(a)| \cdot (b - a)/n$.

41. For a monotone function, the "exact" integral I lies *between* L_n and R_n, i.e., in an interval of
    length $|R_n - L_n| = |f(b) - f(a)| \cdot (b - a)/n$. Since T_n is the *midpoint* of this interval, I
    must lie within *half* the interval's width from T_n. (Draw L_n, R_n, and T_n on a number line to
    understand all this.)

§6.3   Euler's Method: Solving DEs Numerically

1. $y(3) \approx y(2) + y'(2) \cdot 1 = 167.20 - 9.72 = 157.48$.

2. $y(8) \approx y(3) - 8.75 - 7.87 - 7.09 - 6.34 - 5.74 = 121.66$.

3. Differentiation shows that if $y(t) = 70 + 120e^{-0.1t}$, then $y'(t) = -12e^{-0.1t} = -0.1(y - 70)$, as desired.

4. (a) $y'' = -0.1y' = (-0.1)^2(y - 70) > 0$ if $y > 70$.

   (b) Euler's method "follows the tangent line" over each time subinterval. Since $y(t)$ is concave up, it lies above its tangent lines, so the tangent approximation underestimates y.

5. (a) The DE $P' = kP(C - P)$ implies that $P'' = kP'(C - P) - kPP' = kP'(C - 2P)$; this is negative when $2P > C$, or $P > C/2$. That's the case here because $P(0) = 6000 > C/2 = 5000$.

   (b) Since $P(t)$ is concave down and Euler's method is based on tangent lines, Euler's method will overestimate, since tangent lines lie above the graph of $P(t)$.

6. (a) Work quickly shows that given *initial* population 0, the population *remains* at 0. Nothing happens.

   (b) In biological terms, the situation is simply that without any initial breeding members a population can't grow. No parents; no children.

7. (a) $f\big(t_i, Y(t_i)\big) = m_i$.

   (b) $f\big(t_{i+1}, Y(t_{i+1})\big) = m_{i+1}$.

8. (a) It's not hard to guess that the function $y(t) = 3t$ (since $y(0) = 2$, $y(t) = 3t + 2$) solves the IVP above. Thus $y(1) = 5$, $y(2) = 8$, $y(3) = 11$, $y(4) = 14$, $y(5) = 17$.

   (b) In this case, Euler's method gives *exact* values; the Euler estimates commit *no* error.

   (c) Euler's method pretends, in effect, that y' remains constant over small intervals. In this case, y' *is* constant, so Euler's method commits no error. Euler's method will behave this way (i.e., commit *no* error) whenever y' is a constant function.

9. (a) With just 1 subdivision, Euler's method gives

$$e = y(1) \approx y(0) + y'(0) \cdot 1 = 1 + 1 \cdot 1 = 2.$$

This answer *underestimates* e. We can tell this even without knowing the true value of e. The DE $y' = y$ means that $y'(t)$ increases over the interval $0 \le t \le 1$; taking just one Euler step over the interval amounts to pretending that y' is *constant* over this interval.

(b) Carrying out Euler's method with 4 subdivisions, from $t = 0$ to $t = 1$, gives the following table:

step	t	y'	y
0	0.00	1.0000	1.0000
1	0.25	1.2500	1.2500
2	0.50	1.5625	1.5625
3	0.75	1.9531	1.9531
4	1.00	2.4414	2.4414

In particular, $e = y(1) \approx 2.4414$. For the same reason as in the previous part of this exercise, this result *underestimates* the true value of e.

(c) In this case Euler's method underestimate the exact solution for *every n*.

(d) The first few steps show the pattern for $n = 1, 2, 3$:

$$y(1/n) = 1 + 1 \cdot \frac{1}{n}$$

$$y(2/n) = \left(1 + \frac{1}{n}\right) + \left(1 + \frac{1}{n}\right) \cdot \frac{1}{n} = \left(1 + \frac{1}{n}\right)^2$$

$$y(3/n) = \left(1 + \frac{1}{n}\right)^2 + \left(1 + \frac{1}{n}\right)^2 \cdot \frac{1}{n} = \left(1 + \frac{1}{n}\right)^3.$$

The same pattern holds for all n.

10. (a) Results can be tabulated as follows:

t	0	0.4	0.8	1.2	1.6	2.0
y'	1	1	1	1	1	1
y	0	0.4	0.8	1.2	1.6	2.0

They show that the Euler estimate is $Y(2) = 2.0$.

(b) Results can be tabulated as follows:

t	0	0.2	0.4	0.6	0.8	1.0	1.2	1.4	1.6	1.8	2.0
y'	1	1	1	1	1	1	1	1	1	1	1
y	0	0.2	0.4	0.6	0.8	1.0	1.2	1.4	1.6	1.8	2.0

They show that the Euler estimate is $Y(2) = 2.0$.

(c) If $y = t$ then $y' = 1 = 1 + t - y$ and $y(0) = 0$.

(d) In this case, Euler's method produces the exact solution (which could have been predicted, because the slope field shows constant slope 1 along the line $y = t$).

11. (a) Results can be tabulated as follows:

t	0	0.4	0.8	1.2	1.6	2.0
y'	0	0.4	0.64	0.784	0.870	0.922
y	1	1	1.16	1.416	1.730	2.078

They show that the Euler estimate is $Y(2) = 2.078$.

(b) Calculating as in (a), but with step size 0.2, gives the Euler estimate is $Y(2) = 2.107$.

(c) If $y = t + e^{-t}$, then a direct check shows that (i) $y' = 1 - e^{-t} = 1 + t - y$, and (ii) $y(0) = 1$, as required.

(d) The exact value of $y(2)$ is $y(2) = 2 + e^{-2} \approx 2.135$—a little more than both Euler estimates. This underestimate could have been predicted from the slope field, which shows that solutions are concave upward.

12. (a) Results can be tabulated as follows:

t	0	0.4	0.8	1.2	1.6	2.0
y'	2.	1.6	1.36	1.216	1.130	1.078
y	−1.	−0.2	0.44	0.984	1.470	1.922

They show that the Euler estimate is $Y(2) = 1.922$.

(b) Calculating as in (a), but with step size 0.2, gives the Euler estimate is $Y(2) = 1.893$.

(c) If $y = t - e^{-t}$, then a direct check shows that (i) $y' = 1 + e^{-t} = 1 + t - y$, and (ii) $y(0) = -1$, as required.

(d) The exact value of $y(2)$ is $y(2) = 2 - e^{-2} \approx 1.865$—a little less than both Euler estimates. This overestimate could have been predicted from the slope field, which shows that solutions are concave downward in the region in question.

13. (a) $y(1) \approx -0.59374$

(b) $y(1) \approx -0.65330$

(c) $y'(t) = -e^{t} = (2 - e^{t}) - 2 = y(t) - 2$ and $y(0) = 1$

(d) $y(1) = 2 - e \approx -0.71828$.

14. (a) $y(1) \approx 1$

(b) $y(1) \approx 1.25$

(c) $y(1) \approx 1.4194$

(d) $y(1) \approx 1.5240$

(e) $y'(x) = xe^{x^2/2} = xy(x)$ and $y(0) = 1$

(f) $y(1) = e^{1/2} \approx 1.64872$. Thus, the approximation error is (approximately) halved when the number of steps is doubled. This implies that the error made is proportional to $1/n$.

15. (a) $y(0.8) \approx 2.6764$

(b) $y'(t) = (1 - t)^{-2} = \big(y(t)\big)^2$ and $y(0) = 1$

(c) The derivative is changing rapidly and is becoming large. Thus, very small steps are required to achieve an accurate result.

16. (a) If $y(t) = 1 - \cos t$, $y'(t) = \sin t$ and $y(0) = 1 - 1 = 0$. Thus, y is the exact solution of the IVP.

(b) A table helps keep track of results:

step	t	y'	y
0	0.00	0	0
1	0.25	0.2474	0
2	0.50	0.4794	0.0619
3	0.75	0.6816	0.1817
4	1.00	0.8415	0.3521

(c) To plot $y(t)$ and $Y(t)$ on one pair of axes, we use $1 - \cos t$ for y and connect the dots for Y.

(d) $Y(t)$ is the approximation obtained when a left sum is used to estimate the integral in the expression for the exact solution $y(t) = y(0) + \int_0^t y'(x)\,dx$. Since y' is increasing on the interval $[0, 1]$, the left sum approximation underestimates the exact value of the integral. Thus, $Y(t) \le y(t)$.

17. (a) If $y = 1 - e^{-t}$, then a direct check shows that (i) $y'(t) = e^{-t}$ and (ii) $y(0) = 0$, as desired.

(b) Results can be tabulated as follows:

t	0	0.2	0.4	0.6	0.8	1.0
Y'	1	0.819	0.670	0.549	0.449	0.368
Y	0	0.2	0.364	0.498	0.608	0.697

They show that the Euler estimate is $Y(1) \approx 0.697$.

(c) Is a plot.

(d) Since y' is a decreasing function, any left sum estimate of $\int_0^T y'(t)\,dt$ will overestimate the exact answer.

18. (a) The fundamental theorem of calculus says that $y'(t) = f(t)$; if $t = t_0$ in the integral the result is clearly 0.

(b) Euler's method and the left rule lead to identical calculations in this case.

7   *Using the Integral*

§7.1   Measurement and the Definite Integral; Arclength

1.   (b) The area is approximately $L_5 = \dfrac{1}{5} \displaystyle\sum_{i=0}^{4} \left(\dfrac{i}{5} - \dfrac{i^2}{25} \right) = \dfrac{4}{25} = 0.16$.

   (c) The left sum approximation of the area under the curve $y = x - x^2$ *underestimates* the actual area under the curve.

   (d) The area is $\displaystyle\int_0^1 (x - x^2)\, dx = \dfrac{1}{2}x^2 - \dfrac{1}{3}x^3 \Big]_0^1 = \dfrac{1}{6}$.

2.   (b) The exact area is $\displaystyle\int_1^3 (\sin x - \cos x)\, dx$; for this integral, $M_5 \approx 2.2456$.

   (c) The integrand $\sin x - \cos x$ is concave down for all x in $[1, 3]$, the interval of integration, so the midpoint rule overestimates the integral.

   (d) $\text{Area} = \displaystyle\int_1^3 (\sin x - \cos x)\, dx = (-\cos x - \sin x)\Big|_1^3 \approx 2.2306$.

3.   (a) Note $x = 1 - y^2$ implies $y = \sqrt{1 - x}$, so the area is $\displaystyle\int_0^1 \left(1 - \sqrt{1 - x} \right) dx = 1/3$.

   (b) The region is symmetric with that of (a), so its area is also $1/3$.

   (c) Note $x = y + 2$ implies $y = x - 2$, so the area is
   $$\int_1^3 \left(1 - (x - 2) \right) dx = \int_1^3 (3 - x)\, dx = 3x - \dfrac{1}{2}x^2 \Big]_1^3 = 2.$$

   (d) (a) + (b) + (c) = $8/3$.

4.   (a) The area is $\displaystyle\int_0^1 2\sqrt{1 - x}\, dx = \int_{-1}^1 (1 - y^2)\, dy = \dfrac{4}{3}$.

   (b) The area of the triangular region is
   $$\int_1^3 ((2 - x) - (-1))\, dx = \int_{-1}^1 \left(3 - (y + 2) \right) dy = 2.$$

   (c) The region is a square with area 6.

   (d) The area of Region 2 is (c) − (b) − (a) = $6 - 2 - 4/3 = 8/3$.

5.   (a) Applying T_{20} to the given integral I gives $T_{20} \approx 1.47932$.

   (b) Because the integrand is concave up, all midpoint sums underestimate and trapezoid sums overestimate.

6. (a) If $f(x) = x^2 + 1$, then $f'(x) = 2x$, so we want the integral

$I = \int_0^1 \sqrt{1 + f'(x)^2}\, dx = \int_0^1 \sqrt{1 + 4x^2}\, dx$. We estimated this same integral in Example 8, where we got $M_{20} \approx 1.479$.

   (b) Applying the table of integrals gives

$$I = \int_0^1 \sqrt{1 + 4x^2}\, dx = x\sqrt{x^2 + 1/4} + \frac{\ln(x + \sqrt{x^2 + 1/4})}{4}\Bigg]_0^1$$

$$= \frac{\sqrt{5}}{2} + \frac{\ln(2 + \sqrt{5})}{4} \approx 1.4789.$$

7. Because $dy/dx = \cos x$, the length integral is $I = \int_0^\pi \sqrt{1 + \cos^2 x}\, dx$. For this integral, $M_{20} \approx 3.8202$.

8. The length of the line segment is $\int_A^B \sqrt{1 + m^2}\, dx = (B - A)\sqrt{1 + m^2}$. We'd get the same answer using the distance formula, since the line in question joins the points $(A, mA + b)$ and $(B, mB + b)$.

9. The integral has the same *numerical* value in each part:

$\int_0^3 f(t)\, dt = \int_0^3 \left(5t^2 - 20t + 50\right)\, dt = 105$. The parts differ from each other only in interpretation.

   (a) The car traveled 105 miles during the three hours from midnight to 3 am.

   (b) The car traveled 105 feet in the first 3 seconds after midnight.

   (c) The car's velocity increased by 105 feet per minute over the first 3 minutes after midnight.

10. (a) $I = \int_1^3 f(t)\, dt$ represents the object's upward displacement, in feet, from $t = 1$ to $t = 3$. If $I < 0$, the object's net displacement is downward.

   (b) $I = \int_1^3 f(t)\, dt$ represents the object's change in upward velocity, in feet per second, from $t = 1$ to $t = 3$. If $I < 0$, then the object's upward velocity decreased over the interval.

   (c) $I = \int_1^3 f(t)\, dt$ represents the net change in water in the tank over the two-minute interval. If $I < 0$, then the net change in water in the tank decreased.

   $f(t)$ is the net rate of flow of water into a tank, in liters per minute, at time t minutes.

   (d) In this case, $\int_1^3 f(t)\, dt = \int_1^3 g'(t)\, dt = g(t)\Big]_1^3 = g(3) - g(1)$. Thus, if the value of the integral is negative, $g(3) < g(1)$.

11. (a) Estimates will vary depending on which difference quotients are used. The following estimates are computed using "symmetric" difference quotients—e.g., $f'(1.0) \approx (f(1.1) - f(0.9))/0.2$:

t	1.0	1.1	1.2	1.3	1.4	1.5
$f'(t)$	−0.83	−1.17	−1.46	−1.70	−1.875	−1.965
t	1.6	1.7	1.8	1.9	2.0	
$f'(t)$	−1.985	−1.925	−1.78	−1.57	−1.3	

   (b) Different estimates are possible using the data from (a). For example, $M_5 = -1.666$. Here $f(2) - f(1) = -1.666$, too.

   (c) Different numerical integral estimates are possible using the data from the table. For example, $M_5 = 0.1194$. This number estimates the signed area determined by the f-graph.

   (d) Data from the table in (a) can be used to estimate the integral; in this case $M_5 = 1.9494$. This integral gives the length of the f-graph from $t = 1$ to $t = 2$.

   (e) The integrals in question are $I_1 = \int_1^2 2\cos(2t)\,dt$ and $I_2 = \int_1^2 \sin(2t)\,dt$. Evaluating them exactly gives $I_2 = \cos(2)/2 - \cos(4)/2 \approx 0.1187$ and $I_1 = \sin(4) - \sin(2) \approx -1.6661$.

12. (a) $I = \int_0^1 \sqrt{1+x}\,dx = \dfrac{2}{3}(1+x)^{3/2}\Big]_0^1 = \dfrac{4\sqrt{2}-2}{3} \approx 1.219$.

   (b) I is the area under the curve $y = \sqrt{1+x}$ from $x = 0$ to $x = 1$. (The region in question is approximately trapezoidal, with base 1 and altitudes 1 and $\sqrt{2}$.)

   (c) We want a function f for which $I = \int_0^1 \sqrt{1+x}\,dx = \int_0^1 \sqrt{1 + f'(x)^2}\,dx$. We look, therefore, for an f for which $f'(x)^2 = x$, or $f'(x) = \sqrt{x}$. *Any* antiderivative of $\sqrt{x}$ will do; let's use $f(x) = 2x^{3/2}/3$. Plotting this f over $[0, 1]$ gives a graph whose length looks to be about 1.2, as the previous part suggests.

13. Area $= \displaystyle\int_{-1}^1 (1 - x^4)\,dx = \dfrac{8}{5}$.

14. Area $= \displaystyle\int_{-1}^0 (x^3 - x)\,dx + \int_0^1 (x - x^3)\,dx = 2\int_0^1 (x - x^3)\,dx = \dfrac{1}{2}$.

15. Area $= \displaystyle\int_0^1 (x^2 - x^3)\,dx = \dfrac{1}{12}$.

16. Area $= \displaystyle\int_{-1}^2 [(x + 1) - (x^2 - 1)]\,dx = \dfrac{9}{2}$.

17. Area $= \displaystyle\int_0^4 \sqrt{x}\, dx = \dfrac{16}{3}$.

18. Area $= \displaystyle\int_0^1 (\sqrt{x} - x^2)\, dx = \dfrac{1}{3}$.

19. Area $= \displaystyle\int_{-\frac{1}{4}}^1 \left[(2 - x) - \dfrac{9}{4x + 5}\right] dx = \dfrac{65}{32} - \dfrac{9}{4}\ln(9) + \dfrac{9}{4}\ln(4) \approx 0.20666$.

20. Area $= \displaystyle\int_{-1}^1 \left[(2 - x^2) - \dfrac{9}{4x^2 + 5}\right] dx = \dfrac{10}{3} - \dfrac{9}{\sqrt{5}}\arctan\left(\dfrac{2}{\sqrt{5}}\right) \approx 0.39624$.

21. Area $= \displaystyle\int_0^1 e^x\, dx = e - 1$.

22. Area $= \displaystyle\int_{-1}^0 (2^x - 5^x)\, dx + \int_0^1 (5^x - 2^x)\, dx = \dfrac{16}{5\ln 5} - \dfrac{1}{2\ln 2} \approx 1.2669$.

23. The curves intersect at the points $(0, 2)$ and $(5, -3)$. The area is found in two parts, first from $x = -4$ to $x = 0$ and then from $x = 0$ to $x = 5$. The result is
$$\text{Area} = 2\int_{-4}^0 \sqrt{4 + x}\, dx + \int_0^5 (2 - x + \sqrt{4 + x})\, dx = \dfrac{32}{3} + \dfrac{61}{6} = \dfrac{125}{6}.$$

24. The curves intersect at the points $(-7, 4)$ and $(0, -3)$. The area is found in two parts, first from $x = -7$ to $x = 0$ and then from $x = 0$ to $x = 9$. The result is
$$\text{Area} = \int_{-7}^0 \left(\sqrt{9 - x} - (-3 - x)\right) dx + 2\int_0^9 \sqrt{9 - x}\, dx = 36 + \dfrac{127}{6} = \dfrac{343}{6}.$$

25. Note that $y = x^4$ implies $x = \pm y^{1/4}$. Thus Area $= 2\displaystyle\int_0^1 y^{1/4}\, dy = \dfrac{8}{5}$.

26. Note that $y = x^3 \implies x = y^{1/3}$. This and the symmetry of the region give
$$\text{Area} = 2\int_0^1 (y^{1/3} - y)\, dy = \dfrac{1}{2}.$$

27. Note that $y = x^3 \implies x = \sqrt{y}$ and $y = x^3 \implies x = y^{1/3}$. Thus,
$$\text{Area} = \int_0^1 (y^{1/3} - y^{1/2})\, dy = \dfrac{1}{12}.$$

28. Note that $y = x^2 - 1 \implies x = \pm\sqrt{y + 1}$; also, $y = x + 1 \implies x = y - 1$. The graphs cross at $(-1, 0)$ and $(2, 3)$. Thus,
$$\text{Area} = 2\int_{-1}^0 \sqrt{y + 1}\, dy + \int_0^3 (\sqrt{y + 1} - (y - 1))\, dy = \dfrac{4}{3} + \dfrac{19}{6} = \dfrac{9}{2}.$$

29. Note that $y = \sqrt{x} \implies x = y^2$. Thus, Area $= \displaystyle\int_0^2 (4 - y^2)\, dy = \dfrac{16}{3}$.

30. The curves have equations $x = y^2$ and $x = \pm\sqrt{y}$, so we have Area $= \displaystyle\int_0^1 (\sqrt{y} - y^2)\,dy = \frac{1}{3}$.

31. The curves have equations $x = \dfrac{9}{4y} - \dfrac{5}{4}$ and $x = 2 - y$; they cross at the points $(1, 1)$ and

    $(-1/4, 9/4)$. Thus, Area $= \displaystyle\int_1^{9/4} \left(2 - y - \frac{9}{4y} + \frac{5}{4} \right) dy = \frac{65}{32} - \frac{9}{4}\ln\frac{9}{4} \approx 0.2067$.

32. Area $= 10/3 - 9\arctan(2/\sqrt{5})/\sqrt{5}$.

33. The bounding "curves" have equations $x = \ln y$ and $x = 1$. Thus,

    Area $= \displaystyle\int_0^1 (1 - 0)\,dy + \int_1^e (1 - \ln y)\,dy = 1 + (e - 2) = e - 1$.

34. The curves in question have equations $x = \ln(y)/\ln(5)$, $x = \ln(y)/\ln(2)$, $x = -1$, and $x = 1$. Looking at the graphs shows that measuring the region requires four separate integrals: Area $=$

    $$\int_{0.2}^{0.5} \left(\frac{\ln y}{\ln 5} + 1 \right) dy + \int_{0.5}^{1} \left(\frac{\ln y}{\ln 5} - \frac{\ln y}{\ln 2} \right) dy + \int_{1}^{2} \left(\frac{\ln y}{\ln 2} - \frac{\ln y}{\ln 5} \right) dy + \int_{2}^{5} \left(1 - \frac{\ln y}{\ln 5} \right) dy.$$

    The result is messy, but works out decimally to about 1.2669.

35. The curves intersect at $y = -3$ and $y = 2$, so

    $$\text{Area} = \int_{-3}^{2} [(2 - y) - (y^2 - 4)]\,dy$$
    $$= \int_{-4}^{0} 2\sqrt{x + 4}\,dx + \int_{0}^{5} \left[(2 - x) + \sqrt{x + 4} \right] dx = \frac{125}{6}.$$

36. The curves intersect at $y = -3$ and $y = 4$, so Area $= \displaystyle\int_{-3}^{4} [(9 - y^2) - (-3 - y)]\,dy = \frac{343}{6}$.

37. Let the pizza be defined by the circle $x^2 + y^2 = 36$. Then the middle "third" has area

    $4\displaystyle\int_0^2 \sqrt{36 - x^2}\,dx \approx 47.096$ square inches, while the other pieces have area

    $2\displaystyle\int_2^6 \sqrt{36 - x^2}\,dx \approx 33.001$ square inches. The ratio of areas is $47.096/33.001 \approx 1.4271$.

38. (a) If $g(x) = f(x) + C$, for any constant C, then $g'(x) = f'(x)$, so the length of the

    g-graph from $x = a$ to $x = b$ is $\displaystyle\int_a^b \sqrt{1 + g'(x)^2}\,dx = \int_a^b \sqrt{1 + f'(x)^2}\,dx$. The last

    quantity is independent of C.

    (b) Geometrically, the idea is that adding a constant C to f raises or lowers the graph of f, but *doesn't change its length*.

39. length $= \int_a^b \sqrt{1 + f'(x)^2}\, dx = \int_{-1/2}^{\sqrt{3}/2} \frac{1}{\sqrt{1-x^2}}\, dx = \arcsin(\sqrt{3}/2) - \arcsin(-1/2) = \frac{\pi}{2}$.

40. length $= \int_a^b \sqrt{1 + f'(x)^2}\, dx = \int_0^1 \frac{4}{\sqrt{4-x^2}}\, dx = \frac{\pi}{3}$.

41. length $= \int_1^2 \sqrt{1 + (2x)^2}\, dx = \frac{1}{4}\left(4\sqrt{17} + \ln\left(4 + \sqrt{17} \right) - 2\sqrt{5} - \ln\left(2 + \sqrt{5} \right) \right)$
$$\approx 3.1678.$$

42. length $= \int_0^1 \sqrt{1 + e^{2x}}\, dx = \frac{1}{2}\int_1^{e^2} \frac{\sqrt{1+u}}{u}\, du = \frac{1}{2}\int_1^{e^2} \frac{1+u}{u\sqrt{1+u}}\, du$
$$= \left(\sqrt{1+u} + \frac{1}{2}\ln\left| \frac{\sqrt{1+u}-1}{\sqrt{1+u}+1} \right| \right)\Bigg|_1^{e^2} \approx 2.003497110.$$

43. length $= \int_a^b \sqrt{1 + f'(x)^2}\, dx = \int_1^4 \sqrt{1 + \frac{9x}{4}}\, dx = (80\sqrt{10} - 13\sqrt{13})/27 \approx 7.63371$.

44. length $= \int_a^b \sqrt{1 + f'(x)^2}\, dx = \int_0^8 \sqrt{1 + \frac{4}{9x^{\frac{2}{3}}}}\, dx \approx 9.07342$.

45. Using the substitution $u = \sqrt{x}$ we get
$$\text{length} = \int_a^b \sqrt{1 + f'(x)^2}\, dx = \int_1^{16} \sqrt{1 + \sqrt{x}}\, dx = \frac{-8\sqrt{2}}{15} + \frac{40\sqrt{5}}{3} \approx 29.06.$$

46. length $= \int_a^b \sqrt{1 + f'(x)^2}\, dx = \frac{1}{2}\int_{\frac{1}{4}}^4 \sqrt{4 + x^{-6} + \frac{2}{x^4} + x^{-2}}\, dx = 7.76782$.

47. length $= \int_a^b \sqrt{1 + f'(x)^2}\, dx = \int_3^9 \sqrt{\frac{1}{2} + \frac{1}{16x^2} + x^2}\, dx = 36 + \frac{\ln(3)}{4} \approx 36.2747$.

48. length $= \int_a^b \sqrt{1 + f'(x)^2}\, dx = \int_{1/2}^2 \frac{\sqrt{2 + x^{-4} + x^4}}{2}\, dx = \frac{33}{16}$.

49. length $= \int_a^b \sqrt{1 + f'(x)^2}\, dx = \int_{1/2}^2 \sqrt{\frac{1}{2} + \frac{1}{16x^4} + x^4}\, dx = \int_{1/2}^2 \left(x^2 + \frac{1}{4x^2} \right)\, dx = 3$.

50. length $= \int_a^b \sqrt{1 + f'(x)^2}\, dx = \int_1^3 \sqrt{\frac{1}{2} + \frac{1}{16x^6} + x^6}\, dx = \int_3^1 \left(x^3 + \frac{1}{4x^3} \right)\, dx = \frac{181}{9}$.

51. length $= \int_a^b \sqrt{1 + f'(x)^2}\, dx = \int_1^3 \sqrt{\frac{1}{2} + \frac{1}{x^6} + \frac{x^6}{16}}\, dx = \int_1^3 \left(\frac{1}{x^3} + \frac{x^3}{4} \right)\, dx = \frac{49}{9}$.

52. length $= \int_a^b \sqrt{1 + f'(x)^2}\, dx = \int_2^4 \sqrt{\dfrac{1}{16x^8} + \dfrac{1}{2} + x^8}\, dx$

$$= \int_2^4 \left(\dfrac{1}{4x^4} + x^4 \right) dx = \dfrac{761891}{3840} \approx 198.409.$$

53. length $= \int_a^b \sqrt{1 + f'(x)^2}\, dx = \int_1^4 \sqrt{1 + \sinh^2 x}\, dx$

$$= \int_1^4 \cosh x\, dx = \sinh 4 - \sinh 1 \approx 26.1147.$$

54. $L_1 = \int_0^a \sqrt{1 + f'(x)^2}\, dx; \quad L_2 = \int_0^a \sqrt{1 + 4f'(x)^2}\, dx; \quad L_3 = \int_0^{a/2} \sqrt{1 + 4f'(2x)^2}\, dx;$

substituting $t = 2x$ gives $L_3 = \int_0^a \sqrt{1 + 4f'(t)^2}\dfrac{1}{2}\, dt = \int_0^a \sqrt{1/4 + f'(t)^2}\, dt$. It follows that $L_3 \leq L_1 \leq L_2$.

55. The length of the curve $y = f(x)$ from $x = a$ to $x = b$ is $I = \int_a^b \ell(x)\, dx$, where $\ell(x) = \sqrt{1 + \big(f'(x)\big)^2}$. Now $\ell'(x) = f'(x)f''(x)/\sqrt{1 + \big(f'(x)\big)^2}$. Because $f'(x) > 0$ and $f''(x) < 0$, we see that $\ell'(x) < 0$ for $a \leq x \leq b$. Thus, $\ell(x)$ is decreasing on $[a, b]$; hence, L_n *over*estimates the value of I.

56. (a) $J = \int_0^4 \sqrt{1 + \big(f'(x)\big)^2}\, dx$, so

$$\begin{aligned} L_4 &= \sqrt{1 + \big(f'(0)\big)^2} + \sqrt{1 + \big(f'(1)\big)^2} + \sqrt{1 + \big(f'(2)\big)^2} + \sqrt{1 + \big(f'(3)\big)^2} \\ &= \sqrt{1 + (-9)^2} + \sqrt{1 + (-11/2)^2} + \sqrt{1 + (-3)^2} + \sqrt{1 + (-3/2)^2} \\ &\approx 19.611. \end{aligned}$$

   (b) Let $\ell(x)$ be the integrand in J. Note that $\ell'(x) = f'(x)f''(x)/\sqrt{1 + f'(x)^2} < 0$ for $0 \leq x \leq 4$, because $f'(x) < 0$ and $f''(x) > 0$. Thus, $\ell(x)$ is decreasing on $[0, 4]$, and so L_4 *over*estimates J (i.e., $J < L_4$).

57. After 5 minutes, the bug has crawled 30 feet. Since the length of the curve $y = \frac{1}{3}\big(x^2 + 2\big)^{3/2}$ between $x = 1$ and $x = s$ is $s^3/3 + s - 4/3$, the x-coordinate of the bug's position after 5 minutes is $x \approx 4.3271$. Thus, the bug is (approximately) at the point $(4.3271, 31.4470)$.

58. The length of the curve is about 7.6337 feet; at 3 ft/hour the snail takes about 2.54 hours to travel the curve.

59. The fundamental theorem of calculus (FTC) says that $f'(x) = \sqrt{g'(t)^2 - 1}$. Thus $\sqrt{1 + f'(x)^2} = g'(t)$, and the desired arclength is (using the FTC again) $I = \int_a^b g'(t)\, dt = g(b) - g(a)$.

60. A careful algebraic calculation (using the fact that $f'(x)g'(x) = -1$) shows that $1 + h'(x)^2 = (f'(x) - g'(x))^2/4$. Taking square roots of both sides and integrating gives the desired equation.

§7.2 Finding Volumes by Integration

1. Nick = Rick · $\pi/4 \approx 12.47$.

2. $(1.03 + 0.54 \times 10) = 15.7$.

3. The cone has bottom circumference 4 meters, so bottom radius is $2/\pi$. Also, the entire cone has height 40 meters, so its volume is $\pi r^2 h/3 = \pi \cdot 4/\pi^2 \cdot 40/3 = 16.98$ cubic meters. The log part is the bottom half of the cone, which has volume 7/8 of entire cone, or about 14.85 cubic meters.

4. (a) The volume of a circular cylinder is $\pi r^2 h$; the Fact implies that a cone has 1/3 this volume.

    (b) Let $A(y)$ be the cross-sectional area of the cone at height h. The radius at height y decreases linearly from r (when $y = 0$) to 0 (when $y = h$). This radius is therefore $r(1 - y/h)$, and so the cross-sectional area at height y is $\pi r^2(1 - y/h)^2$. Integrating gives $\int_0^h \left(\pi r^2(1 - y/h)^2 \right) dy = \pi r^2 \int_0^h (1 - y/h)^2 \, dy = \dfrac{\pi r^2 h}{3}$.

5. (a) The volume of a square cylinder with side s and height h is $s^2 h$; the Fact implies that a pyramid has 1/3 this volume.

    (b) Let $A(y)$ be the cross-sectional (square) area of the pyramid at height h. The side length at height y decreases linearly from s (when $y = 0$) to 0 (when $y = h$). This radius is therefore $s(1 - y/h)$, and so the cross-sectional area at height y is $s^2(1 - y/h)^2$. Integrating gives $\int_0^h s^2(1 - y/h)^2 \, dy = s^2 \int_0^h (1 - y/h)^2 \, dy = \dfrac{s^2 h}{3}$.

6. (a) If we cut off the top half of a cone, then the new solid has base radius $r/2$ and height $h/2$. The cone volume formula implies that the new cone has volume $\pi r^2 h/24$, which is 1/8 of the old volume.

    (b) If we cut off the top half of a square pyramid, then the new solid has base edge $s/2$ and height $h/2$. The pyramid volume formula implies that the new cone has volume $s^2 h/24$, which is 1/8 of the old volume.

    (c) If we cut off the top half of any hat solid, then the new solid has base area 1/4 of the old base area, and the new height is 1/2 the old height. Thus the new volume is 1/8 the old volume.

7. The cone's cross sectional area at height y is $\pi r^2(1 - y/h)^2$. Since the cone's total volume is $\pi r^2 h/3$, we want H so that $\pi r^2 \int_0^H (1 - y/h)^2 \, dy = \pi r^2 h/6$. Solving this equation for H gives $h(1 - 1/\sqrt[3]{2})$.

8. The pyramid's cross sectional area at height y is $s^2(1 - y/h)^2$. Since the cone's total volume is $s^2 h/3$, we want H so that $s^2 \int_0^H (1 - y/h)^2 \, dy = s^2 h/6$. Solving this equation for H gives $h(1 - 1/\sqrt[3]{2})$.

9. (a) The given sum is a Riemann sum for the volume integral $V = \int_a^b \pi f(x)^2\, dx$.

(b) If $f(x)$ is decreasing, then so is the integrand $\pi f(x)^2$ in the volume integral. Therefore, all right sums underestimate.

(c) If $f(x) = x^2$, $a = 0$, and $b = 2$, then the given sums converge to the integral $V = \int_0^2 \pi x^4\, dx$, which has value $32\pi/5$.

10. (a) The given sum is a Riemann sum for the volume integral $V = \int_a^b \pi g(y)^2\, dy$.

(b) If $g(y)$ is decreasing, then so is the integrand $\pi g(y)^2$ in the volume integral. Therefore, all left sums overestimate.

(c) If $g(y) = \sqrt{y}$, $a = 0$, and $b = 4$, then the given sums converge to the integral $V = \int_0^4 \pi y\, dy$, which has value 8π.

11. $V = \displaystyle\int_0^8 \pi \left(x^3\right)^2 dx = \dfrac{8^7 \pi}{7}$.

12. $V = \displaystyle\int_0^1 \pi \left(1^2 - (x^4)^2\right) dx = \dfrac{8\pi}{9}$.

13. $V = \displaystyle\int_0^2 \pi \left((x+6)^2 - (x^3)^2\right) dx = \dfrac{1688\pi}{21}$.

14. $V = \displaystyle\int_0^1 \pi \left((x^2)^2 - (x^3)^2\right) dx = \dfrac{2\pi}{35}$.

15. $V = \displaystyle\int_0^2 \pi \left(4^2 - (y^2)^2\right) dy = \dfrac{128\pi}{5}$.

16. $V = \displaystyle\int_0^1 \pi \left((\sqrt{y})^2 - (y^2)^2\right) dy = \dfrac{3\pi}{10}$.

17. $V = \displaystyle\int_0^4 \pi \left(\sqrt{y}\right)^2 dy - \int_1^4 \pi \left(\log_2 y\right)^2 dy = \pi \left(\dfrac{16}{\ln 2} - 8 - \dfrac{6}{(\ln 2)^2}\right) \approx 8.15214$.

18. $V = \displaystyle\int_0^e \pi 1^2\, dy - \int_1^e \pi (\ln y)^2\, dy = 2\pi$.

19. For x in $[-2, 2]$, the cross-section at x is a square with edge-length $2\sqrt{4 - x^2}$; this square has area $4(4 - x^2)$. Thus the solid's volume is $4\int_{-2}^2 (4 - x^2)\, dx = 128/3$.

20. For y in $[0, 2]$, the cross-section at y is a semicircle with radius $x = \sqrt{2y}$. This semicircle has area πy, so the solid's volume is $\int_0^2 \pi y\, dy = 2\pi$.

21. Recall that an equilateral triangle of base s has area $\sqrt{3}s^2/4$. For x in $[0, 3]$, the cross-section at x is an equilateral triangle with base $s = 2\sqrt{3x}$. This triangle has area $3\sqrt{3}x$, so the solid's volume is $3\sqrt{3}\int_0^3 x\, dx = 27\sqrt{3}/2$.

22. The area of an isosceles right triangle with hypotenuse h is $h^2/4$. For x in $[1, 4]$, the cross-section at x is an isosceles right triangle with base $1/x$, and hence area $1/(4x^2)$. Thus, the volume of the solid is $\int_1^4 \dfrac{dx}{4x^2} = \dfrac{3}{16}$.

23. (a) Circular cross sections all have area πr^2, so the volume is $\int_0^h \pi r^2\, dx = \pi r^2 h$.

   (b) Think of the cylinder as having its base inside the circle $x^2 + y^2 = r^2$ in the xy-plane. Cross sections perpendicular to the x-axis are then rectangles, with base $2\sqrt{r^2 - x^2}$ and height h, for $-r \le x \le r$. Thus the volume in question is $\int_{-r}^r 2\sqrt{r^2 - x^2}h\, dx = \pi r^2 h$.

24. (a) Suppose that the base of the right circular cylinder is the circle $x^2 + y^2 = r^2$. Since the wedge is cut at an angle of $45°$, cross-sections of the wedge cut perpendicular to the y-axis are isosceles right triangles whose base and height are $\sqrt{r^2 - y^2}$. Thus, the area of the cross-section at y is $A(y) = (r^2 - y^2)/2$ so the volume of the wedge is
$$V = \int_{-r}^r A(y)\, dy = 2r^3/3.$$

   (b) Suppose that the base of the right circular cylinder is the circle $x^2 + y^2 = r^2$. Since the wedge is cut at an angle of $45°$, cross-sections of the wedge cut perpendicular to the x-axis are rectangles with base $2\sqrt{r^2 - x^2}$ and height $r - x$. Thus, the area of the cross-section of x is $A(x) = 2(r - x)\sqrt{r^2 - x^2}$ so the volume of the wedge is
$$V = \int_{-r}^r A(x)\, dx = \pi r^3.$$

25. The radius of the glass (in inches) at height h (in inches) is $r = 1 + 0.1h$ when $0 \le h \le 5$. Thus, the volume of the glass is $V = \pi \int_0^5 (1 + 0.1h)^2\, dh = \dfrac{95\pi}{12} \approx 24.87 \text{ in}^3$.

26. The radius of the dish at height h inches is $r = 4 + h/2$ inches, for $0 \le h \le 2$. Thus, the volume is $V = \pi \int_0^2 (4 + h/2)\, dh = \dfrac{122\pi}{3} \approx 127.76 \text{ in}^3$.

27. From the information given, the radius of the Earth is $r = C/2\pi \approx 3,963$ miles.

   (a) $V = \pi \int_{\sqrt{2}r/2}^r (r^2 - y^2)\, dy = \left(\dfrac{2}{3} - \dfrac{5\sqrt{2}}{12}\right)\pi r^3 \approx 1.1514 \times 10^{10} \text{ miles}^3$.

   (b) $V = \pi \int_0^{\sqrt{2}r/2} (r^2 - y^2)\, dy = \dfrac{5\sqrt{2}\pi r^3}{12} \approx 1.1522 \times 10^{11} \text{ miles}^3$.

28. The volume of water in the balloon is $V = \pi \int_{-3}^1 (9 - y^2)\, dy = 80\pi/3$ cubic inches.

29. **Method 1:** Cross-sections parallel to the ends of the tank have constant area:
$$A(x) = 2\int_{-4}^2 \sqrt{4^2 - y^2}\, dy = 2\left(\frac{1}{2}y\sqrt{4^2 - y^2} + 8\arcsin\left(\tfrac{y}{4}\right)\right)\Bigg]_{-4}^2 = 4\sqrt{3} + \frac{32\pi}{3}.$$

Thus, the volume of gasoline in the tank is $V = \int_0^{25} A(x)\,dx = 100\sqrt{3} + \dfrac{800\pi}{3} \approx 1011.0$ cubic feet.

Method 2: The cross-section parallel to the surface at height y has area $A(y) = 25 \cdot 2\sqrt{4^2 - y^2}$. Thus, the volume of gasoline in the tank is

$$V = \int_{-4}^{2} A(y)\,dy = 50 \int_{-4}^{2} \sqrt{4^2 - y^2}\,dy.$$

Method 3: The volume of the tank is 400π cubic feet. The volume of the "empty" portion at the top is $25 \int_{-\sqrt{12}}^{\sqrt{12}} \left(\sqrt{4^2 - x^2} - 2 \right) dx$ cubic feet. Now the volume of the gasoline in the tank can be found by subtraction.

30. Assume that the bowl is the lower portion of a sphere of radius r (i.e., the portion from $y = -r$ to $y = 0$). The liquid remaining in the bowl after it has been tipped fills the portion of the bowl from $y = -r$ to $y = -r\sqrt{2}/2$. Thus, the volume of the liquid remaining in the bowl is

$$V = \int_{-r}^{-r\sqrt{2}/2} \pi (r^2 - y^2)\,dy = \pi r^3 \left(\frac{2}{3} - \frac{5\sqrt{2}}{12} \right).$$

31. The volume is the integral $\int_0^{30} A(h)\,dh$, where $A(h)$ is the leg's cross-sectional area at height h; these areas can be found by squaring the width at each height. Thus we can find $A(0), A(5), \ldots A(30)$. Using these values gives the integral approximations $L_6 = 39.8$ $R_6 = 27.8$; $T_6 = 33.8$; $M_3 = 33.2$, all in cubic inches. (Alternatively, one might use the linear formula $w = 1.6 - 1.2h/30$ for width as a function of height; squaring this gives a quadratic formula for area, which can be integrated.)

32. The cross-sectional area at height h is $V'(h)$. Thus, the cross-sectional area of the object 1 inch above its base is $1.5 + \cos 1$ square inches.

33. Cross-sections parallel to the ends of the tank have area:

$$A(x) = 2 \int_{-6}^{3} \sqrt{9 - \frac{y^2}{4}}\,dy = 4 \int_{-3}^{3/2} \sqrt{9 - u^2}\,du$$

$$= 2 \left(u\sqrt{9 - u^2} + 9 \arcsin \left(\tfrac{u}{3} \right) \right) \Big|_{-3}^{3/2} = \frac{9\sqrt{3}}{2} + 12\pi.$$

Thus, the volume of fuel oil in the tank is $V = \int_0^{10} A(x)\,dx = 45\sqrt{3} + 120\pi \approx 454.93$ cubic feet.

34. Imagine the two cylinders as having the x-axis and the y-axis as their axes; a third axis, the z-axis, sticks straight up. The solid is the intersection of these two cylinders. Cross-sections of this solid perpendicular to the z-axis are squares; the cross section at height z is a square

with side $2\sqrt{R^2 - z^2}$; the area of this square is $4(R^2 - z^2)$. Thus the solid's volume is

$$\int_{-R}^{R} 4(R^2 - z^2)\, dz = \frac{16R^3}{3}.$$ (Note that this volume is $4/\pi$ times the volume of the sphere of radius R.)

35. (a) The rectangle has base $(b - a)$ and height h, so the area is $(b - a)h$.

   (b) Rotating R about the x-axis produces a circular cylinder of radius h and length $(b - a)$, so the volume is $\pi h^2(b - a)$.

   (c) Rotating R about the line $y = -c$ produces a hollow tube with length $(b - a)$, outer radius $c + h$, and inner radius c. Thus the volume is the
   $$\pi(c + h)^2(b - a) - \pi c^2(b - a) = \pi(2hc + h^2)(b - a).$$

   (d) Rotating R about the line $y = c$, where $c \geq h$, produces a hollow tube with length $(b - a)$, outer radius c, and inner radius $c - h$. Thus the volume is the
   $$\pi c^2(b - a) - \pi(c - h)^2(b - a) = \pi(2hc - h^2)(b - a).$$

36. (a) $\pi \displaystyle\int_0^4 \left((6 - y) - (-2)\right)^2 dy - \pi \int_0^4 \left(\sqrt{y} - (-2)\right)^2 dy.$

   (b) $\pi \displaystyle\int_0^2 \left(x^2 - (-1)\right)^2 dx + \pi \int_2^6 \left((6 - x) - (-1)\right)^2 dx.$

37. $V = \displaystyle\int_0^1 \pi \left(1^2 - \left(1 - \sqrt{x}\right)^2\right) dx = \frac{5\pi}{6}.$

38. $V = \displaystyle\int_{-1}^2 \pi \left(\left((x + 1) + 1\right)^2 - \left((x^2 - 1) + 1\right)^2\right) dx = \frac{72\pi}{5}.$

39. $V = \displaystyle\int_{-4}^0 \pi \left(\left(2 + \sqrt{x + 4}\right)^2 - \left(2 - \sqrt{x + 4}\right)^2\right) dx$

   $\qquad + \displaystyle\int_0^5 \pi \left(\left(2 + \sqrt{x + 4}\right)^2 - (2 - (2 - x))^2\right) dx$

   $\qquad = \dfrac{128\pi}{3} + \dfrac{123\pi}{2} = \dfrac{625\pi}{6}.$

40. $V = \displaystyle\int_0^1 \pi \left(\left(\sqrt{x} + 2\right)^2 - \left(x^2 + 2\right)^2\right) dx = \frac{49\pi}{30}.$

41. Note that $y = \arctan x$ implies $x = \tan y$. Thus,

   $$V = \pi \int_0^{\pi/4} \left(1 - \tan^2 y\right) dy = \pi\left(y - (\tan y - y)\right)\Big]_0^{\pi/4} = \pi\left(\frac{\pi}{2} - 1\right) \approx 1.7932.$$

42. (a) $V = \pi \displaystyle\int_{-3}^{-1} \arctan^2 x\, dx.$

(b) For any n, L_n overestimates V since $\left(\arctan^2 x\right)' = \dfrac{2\arctan x}{1+x^2} < 0$ over the interval of integration (i.e., the integrand is a decreasing function).

43. The solid is a "thick-walled cylinder," with outer radius b, inner radius a, and height h. Thus, its volume is $\pi b^2 h - \pi a^2 h$.

44. (a) Each summand in question has the form $f(m_j)\left((x_j)^2 - (x_{j-1})^2\right)$, where m_j is the midpoint of the small interval $[x_j, x_{j-1}]$. Problem #43 shows that this quantity is the volume of the "shell" formed by rotating a thin vertical strip in R around the y-axis. Putting these vertical strips together gives a good approximation to R, so adding the volumes of all the shells approximates the volume obtained by rotating the entire region R about the y-axis.

    (b) For any u and v, we have
    $u^2 - v^2 = (u-v)(u+v) = 2 \cdot (u-v) \cdot (u+v)/2 = 2(u-v)m$, where $m = (u+v)/2$.
    Applying this fact with $u = x_j$, $v = x_{j-1}$, and $m_j = (u+v)/2$ does the trick.

    (c) Parts (a) and (b) say that
    $$V \approx \pi \sum_{j=1}^{n} f(m_j)\left((x_j)^2 - (x_{j-1})^2\right) = \pi \sum_{j=1}^{n} f(m_j)2m_j(x_j - x_{j-1}).$$ The last
    expression is an approximating sum for the integral $2\pi \int_a^b x f(x)\, dx$, which therefore gives the solid's volume, as desired.

45. The volume formula $V = 2\pi \displaystyle\int_a^b x f(x)\, dx$ gives $2\pi \int_0^\pi x \sin x\, dx = 2\pi^2$. (Use the table of integrals to find an antiderivative.)

46. The volume formula $V = 2\pi \displaystyle\int_a^b x f(x)\, dx$ gives $2\pi \int_0^R x\sqrt{R^2 - x^2}\, dx = 2\pi R^3/3$.

47. The shell method gives $V = 2\pi \int_0^2 x(1 - x/2)\, dx = 4\pi/3$. (The problem can also be done in other ways.)

48. The shell method gives $V = 2\pi \int_0^1 x e^{-x^2}\, dx = \pi - \pi/e$.

49. The shell method gives $V = 2\pi \displaystyle\int_1^e x \ln x\, dx = \pi \dfrac{e^2 + 1}{2}$.

50. The original sphere has volume $4\pi r^3/3 = 36\pi$. The part drilled out is the solid formed by revolving about the y-axis the region inside the circle $x^2 + y^2 = 9$ and bounded by the lines $x = 0$ and $x = 2$. The shell method easily gives the volume of the upper half of this region:
    It's $V = 2\pi \displaystyle\int_0^2 x\sqrt{9 - x^2}\, dx = 18\pi - \dfrac{10\sqrt{5}\,\pi}{3}$. The drilled out region has volume
    $36\pi - \frac{20\sqrt{5}}{3}\pi$. The volume remaining is $36\pi - (36\pi - \frac{20\sqrt{5}}{3}\pi) = \frac{20\sqrt{5}}{3}\pi$. Finding this directly using shells gives the volume of the upper half of this region as:
    $V = 2\pi \int_2^3 x\sqrt{9 - x^2}\, dx = 10\sqrt{5}\pi/3$; doubling this gives $20\sqrt{5}\pi/3$.

51. The remaining solid can be formed by revolving the region bounded by $y = 0$, $x = a$, and $y = h - hx/r$ about the y-axis. The shell method gives

$$V = 2\pi \int_a^r xh(1 - x/r)\, dx = -\pi a^{2h} + \frac{2\pi a^3 h}{3r} + \frac{\pi h r^2}{3}.$$

52. (a) $V = \pi \displaystyle\int_0^2 (8^2 - y^6)\, dy = \dfrac{768\pi}{7}.$

    (b) $V = 2\pi \displaystyle\int_0^8 x^{4/3}\, dx = \dfrac{768\pi}{7}.$

53. (a) Solving the equation $y = x\sqrt{1 - x^2}$ for x^2 yields $x^2 = \left(1 \pm \sqrt{1 - 4y^2}\right)/2$. Thus, the outer boundary of the region is the curve $x = \left(\dfrac{1 + \sqrt{1 - 4y^2}}{2}\right)^{1/2}$ and the inner boundary of the region is the curve $x = \left(\dfrac{1 - \sqrt{1 - 4y^2}}{2}\right)^{1/2}$ when $0 \le y \le 1/2$. Therefore, the volume of the solid of revolution is

$$V = \int_0^{1/2} \frac{\pi}{2}\left(1 + \sqrt{1 - 4y^2}\right) dy - \int_0^{1/2} \frac{\pi}{2}\left(1 - \sqrt{1 - 4y^2}\right) dy.$$

    (b) Using the method of cylindrical shells, $V = \displaystyle\int_0^1 \pi x f(x)\, dx = \int_0^1 2\pi x^2 \sqrt{1 - x^2}\, dx.$

    (c) The integral in part (b) can be evaluated using the substitutions $u = 1 - x^2$, $w = u - 1/2$ and the table of integrals:

$$\int_0^1 2\pi x^2 \sqrt{1 - x^2}\, dx = \pi \int_0^1 \sqrt{(1/2)^2 - (u - 1/2)^2}\, du$$

$$= \pi \int_{-1/2}^{1/2} \sqrt{(1/2)^2 - w^2}\, dw = \frac{\pi^2}{8}.$$

54. The integral adds up small contributions of the form $rv(r)\Delta r$; each such contribution gives the rate of flow (in cubic cm per second) through a small ring-shaped subset of the tube (with inner radius r and thickness Δr. The sum of all such contributions—the integral—gives the total flow rate through the entire tube.

55. (a) Let $p(r)$ be the population density at a distance r from the center of the city. Since p is a linear function it can be written in the form $p(r) = ar + b$. Now, $p(0) = K$ and $p(R) = 0$, so $b = K$ and $a = -K/R$. Thus, $p(r) = -Kr/R + K = K(1 - r/R)$.

    (b) The population of the city is $\displaystyle\int_0^R p(r)\, dr = \dfrac{KR}{2}.$

§7.3 Work

1. Let x denote the number of inches of compression.

   (a) To find k, use the equation (implicit in the problem statement) $F(2) = 2k = 10$. It follows that $k = 5$ (and hence that $F(x) = 5x$).

   (b) Compressing from 16 inches to 12 inches means compressing from $x = 2$ to $x = 6$. Thus the work done is

   $$W = \int_2^6 F(x)\, dx = \int_2^6 5x\, dx = 80 \text{ inch-pounds.}$$

2. (a) $W = \int_0^2 F(x)\, dx = \int_0^2 40x\, dx = 80$ ft-lbs.

   (b) $W = \int_0^s F(x)\, dx = \int_0^s 40x\, dx = 20s^2$ ft-lbs.

   (c) By the problem and the previous part, $W = 20s^2 = 10000$. Solving this for s gives $s = 10\sqrt{5} \approx 22.3361$ feet.

3. (a) work $= \int_0^{10} kx\, dx = 50k$ ft-lbs.

   (b) work $= \int_a^{a+10} kx\, dx = k(50 + 10a)$ ft-lbs.

4. In Example 2 the orbit had "altitude" 25,000 km; that corresponds to a distance of 31,000 km miles from earth's center. The current situation differs only in that the orbit is now 30,000 km from earth's center (or, equivalently, at "altitude" 24,000 km). The reasoning in Example 2 shows that the work done (in N-km) in moving from $x = 6000$ to $x = 30000$ is given by the integral:

   $$\int_{6000}^{30,000} \frac{1.44 \times 10^{11}}{x^2} = -\frac{1.44 \times 10^{11}}{x} \Bigg]_{6000}^{30,000} = 19,200,000 \text{ N-km.}$$

   (Compare this to the result of Example 2, which showed that raising the satellite to 31,000 km took 19,350,000 N-km of work. One point of this problem is that the next 1000 km of altitude comes cheaply—it "costs" only 150,000 N-km of work.)

5. work $= \int_0^{200} 5 \cdot (200 - x)\, dx = 100,000$ ft-lbs.

6. Each parallel slice is a circular cylinder with cross-sectional area 25π and thickness Δx. Therefore,

   $$\text{work} = \int_0^5 62.4 \cdot 25\pi \cdot (10 - x)\, dx = 58,500\pi \text{ ft-lbs} \approx 183,783 \text{ ft-lbs.}$$

7. For reasons as in Example 5, the necessary forces F_1 and F_2 for the two buckets are given, respectively, by

$$F_1(x) = 60 + 0.25 \cdot (60 - x) = 75 - \frac{x}{4};$$
$$F_2(x) = 50 + 0.25 \cdot (70 - x) = 67.5 - \frac{x}{4}.$$

(The unit of force is pounds; x measures distance in feet from the bottom of the well.) To find the work in each case, we integrate:

$$W_1 = \int_0^{60} F_1(x)\,dx = \int_0^{60} \left(75 - \frac{x}{4}\right) dx = 4050 \text{ ft-lbs};$$
$$W_2 = \int_0^{70} F_2(x)\,dx = \int_0^{70} \left(67.5 - \frac{x}{4}\right) dx = 4112.5 \text{ ft-lbs}.$$

Raising the *second* bucket takes a little more work.

8. (a) Let $F(x) = ax + bx^2$ be the force. We're given that $F(1) = 15$ and $F(2) = 50$. Thus, $a + b = 15$ and $2a + 4b = 50$; solving gives $a = 5$ and $b = 10$. Thus $F(x) = 5x + 10x^2$.

   (b) Work $= \int_0^1 (5x + 10x^2)\,dx = \frac{35}{6}$ ft-lb.

   (c) Work $= \int_0^2 (5x + 10x^2)\,dx = \frac{110}{3}$ ft-lb.

9. Density ρ means that the fluid's mass is ρ times its volume. To solve the problem we imitate Example 6. Imagine the tank as the solid formed by revolving the lower half of the circle $x^2 + y^2 = r^2$ around the y-axis. At each height y with $-r \le y \le 0$, the thin slab at height y, with thickness Δy, has area $\pi x^2 = \pi(r^2 - y^2)$, volume $\pi(r^2 - y^2)\Delta y$, and mass $\rho\pi(r^2 - y^2)\Delta y$. The force needed to raise the slab is $g\rho\pi(r^2 - y^2)\Delta y$, where g is the gravitational constant. The slab moves through distance $-y$, so the work done on the slab is $-y \cdot g\rho\pi(r^2 - y^2)\Delta y$. Integrating over y gives the answer:

Work $= g\rho\pi \displaystyle\int_{-r}^{0} -y(r^2 - y^2)\,dy = \dfrac{\pi\rho g r^4}{4}$. (The units of the answer depend on the units given for the data.)

10. To solve the problem we imitate Example 6. Imagine the tank as the solid generated by revolving the trapezoid with corners $(0,0)$, $(1,0)$, $(2,1)$, and $(0,1)$ about the y-axis. At height y, the tank's radius is $1 + y$, so its cross-sectional area is $\pi(1 + y)^2$. The slab at height y is lifted a distance $1 - y$. Thus, the work done is

Work $= g\rho\pi \displaystyle\int_0^1 (1 - y)(1 + y)^2\,dy = \dfrac{11\pi\rho g}{12} \approx 28.22\rho$. (The units of work here are newton-meters, or Joules.)

11. work $= \displaystyle\int_{-4}^{4} 42 \cdot 2\sqrt{16 - y^2} \cdot 15 \cdot (y + 17)\, dy$

$\qquad\quad = -420\left(16 - y^2\right)^{3/2} + 10{,}710\, y\sqrt{16 - y^2} + 171{,}360\,\arcsin(y/4)\,\Big|_{-4}^{4}$

$\qquad\quad = 171{,}360\pi \text{ ft-lbs} \approx 538{,}343 \text{ ft-lbs}.$

12. Let's find an expression for $F(x)$, the force required to lift the bucket, water, and rope when the bucket is x feet from the bottom.

Since the bucket loses 40 pounds of water in 75 feet of vertical travel, it loses $40/75 = 8/15$ pounds per foot. Thus, at height x, the bucket and water weigh $80 - 8x/15$ pounds. When the height is x, $75 - x$ feet of rope must be lifted; they weigh $(75 - x) \cdot 0.65$ pounds. Putting it all together:

$$F(x) = 80 - \frac{8x}{15} + (75 - x) \cdot 0.65 \text{ pounds.}$$

Thus the necessary work W is given by

$$W = \int_{0}^{75} F(x)\, dx = \int_{0}^{75} \left(80 - \frac{8x}{15} + (75 - x) \cdot 0.65\right) dx = 6328.125 \text{ ft-lbs.}$$

13. Let x denote the distance (in feet) that the spring is extended. Notice that since the chain weighs 20 pounds, we start with $x = 5$. (Draw a picture! Note, too, that other x-scales are possible.)

The problem is to find $\int_{5}^{7} F(x)\, dx$, where $F(x)$ is the net downward force necessary at a given x.

Let's find a formula for $F(x)$; there are two main ingredients: (i) For any value of x, the spring exerts an *upward* force of $4x$ pounds. (ii) For a given value of x, the length of chain remaining above the floor is $10 - (x - 5) = 15 - x$ feet. (A diagram should make this convincing.) Since the chain weighs 2 pounds per foot, this length of chain exerts a *downward* force of $2(15 - x) = 30 - 2x$ pounds.

Putting (i) and (ii) together means that the *net* downward force required for given x is $F(x) = 4x - (30 - 2x) = 6x - 30$ pounds. Thus the desired work is

$$W = \int_{5}^{7} F(x)\, dx = \int_{5}^{7} (6x - 30)\, dx = 12 \text{ ft-lbs.}$$

14. In each case, the work done is given by the integral $W = \int_{a}^{b} F(x)\, dx = \int_{a}^{b} k f'(x)\, dx$.

(a) If $k = 10$, $a = 0$, $b = 1$, $f(x) = x$, then $W = \displaystyle\int_{0}^{1} 10 \cdot dx = 10 \cdot x \Big]_{0}^{1} = 10.$

(b) If $k = 10$, $a = 0$, $b = 1$, $f(x) = x^3$, then $W = \displaystyle\int_{0}^{1} 10 \cdot 3x^2\, dx = 10 \cdot x^3 \Big]_{0}^{1} = 10.$

(c) If $k = 10$, $a = 0$, $b = 1$, $f(x) = x^n$, and n is *any* positive integer, then
$$W = \int_0^1 10 \cdot nx^{n-1}\, dx = 10 \cdot x^n\Big]_0^1 = 10.$$

(d) We got the same answer each time. Mathematically, the point is that

$$W = \int_a^b F(x)\, dx = \int_a^b kf'(x)\, dx = k\, f(x)\Big]_a^b = k(f(b) - f(a)).$$

Physically, the point is that in every case, the work is always the product of the object's weight and the *vertical* distance through which it travels. In a sense, then, the particular curve along which it travels makes no difference.

15. The work is the signed area under the graph of F from $x = a$ to $x = b$.

16. Work $= \displaystyle\int_a^b F(x)\, dx = \int_a^b \left(\frac{A}{x^7} + \frac{B}{x^{13}}\right) dx = \frac{A}{6a^6} - \frac{A}{6b^6} + \frac{B}{12a^{12}} - \frac{B}{12b^{12}}.$

17. (a) The expression $\displaystyle\int_0^h A(y)\, dy$ represents the volume of the tank.

(b) The slab at height y gets lifted a distance of $h - y$, so the needed work (using foot-pounds as units) is $62.4 \int_0^a A(y)(h - y)\, dy$.

18. (a) The work is greater if the fat end of the cone is down rather than up because in the former case more water is originally lower, and so needs to be pumped higher.

(b) The calculation resembles that in Example 6, except that the radius at height y is $r(1 - y/h)$. Thus the work done is $62.4\pi r^2 \displaystyle\int_0^h (h - y)(1 - y/h)^2\, dy = 62.4\pi r^2 \frac{h^2}{4}.$
(This is three times the work needed in the situation of Example 6.)

§7.4 Separating Variables: Solving DEs Symbolically

1. (a) 200 °F; 160 °F; 120 °F; 70 °F; 40 °F.

    (b) The value of T_0 is the temperature of the coffee at time $t = 0$. Thus, the figure shows the cooling curves for cups of coffee that start at each of the temperatures listed in part (a).

2. $y' = -0.1(y - 70)$, $y(0) = 200$.

3. $y' = -0.1(y - 70)$, $y(0) = 70$.

4. The problem statement implies that $y(10) = 100 = (T_0 - 70)e^{-0.1 \cdot 10} + 70$. Therefore, $T_0 = 30e^1 + 70 \approx 151.55$.

5. (a) One way to find the values of T_r that correspond to C_1, C_4, and C_9 is just to observe that T_r represents the "long-run" temperature—the temperature to which the coffee tends over a long period of time. Looking at the right-hand parts of the graphs lets us read off these numbers for C_1 and C_4. Thus C_1 corresponds to $T_r = 100$; C_4 corresponds to $T_r = 70$. Not enough of C_9 appears for this to work very well. Instead, we can use the fact that for C_9, $y(10) = 70$. From this it follows:

    $$y(10) = (190 - T_r)e^{-1} + T_r = 70 \implies T_r \approx 0.$$

    (b) The value of T_r is the temperature of the room. Thus, the values found in part (a) give the temperature of the room in which each cup of coffee cooled.

6. $y' = -0.1(y - 100)$, $y(0) = 190$.

7. For C_9, $y(10) = 70$. Therefore, $y(10) = (190 - T_r)e^{-1} + T_r = 70 \implies T_r \approx 0$. Thus, the IVP corresponding to curve C_9 is $y' = -0.1y$; $y(0) = 190$.

8. Looking carefully at the picture. A curve that represents coffee that starts at 190 degrees and is at 100 degrees after 20 minutes would fall about *halfway* between C_2 and C_3. In particular, C_2 and C_3 correspond to room temperatures of 90 and 80 degrees, respectively. Thus the desired curve would correspond to room temperature of about 85 degrees.

9. We need to check (1) that $y(t) = (T_0 - T_r)e^{-0.1t} + T_r$ solves the DE $y' = -0.1(y - T_r)$; and (2) that $y(0) = T_0$. The latter is easy: $y(0) = (T_0 - T_r)e^0 + T_r = T_0$, as claimed.

    To check (1), we calculate both sides of the DE and see that they agree:

    $$\begin{aligned} y(t)' &= (T_0 - T_r) \cdot e^{-0.1t} \cdot (-0.1); \\ -0.1(y - T_r) &= (-0.1) \cdot (T_0 - T_r)e^{-0.1t}. \end{aligned}$$

10. If $T_0 = 200$ and $y(10) = 100$, then
    $$y(10) = (200 - T_r)e^{-1} + T_r = 100 \implies T_r = -\frac{200\,e^{-1} - 100}{-e^{-1} + 1} \approx 41.802.$$

11. If $T_r = 80$ and $y(10) = 120$, then
    $$y(10) = (T_0 - 80)e^{-1} + 80 = 120 \implies T_0 = -\left(-\tfrac{80}{e} - 40\right)e \approx 188.731.$$

12. We can use the information $y(10) = 100$ and $y(20) = 80$ to solve for *both* constants T_0 and T_r, as follows:

    $$y(10) = 100 \implies 100 = (T_0 - T_r) \cdot e^{-1} + T_r;$$
    $$y(20) = 80 \implies 80 = (T_0 - T_r) \cdot e^{-2} + T_r.$$

    Solving these two equations gives $T_r \approx 68.36$, $T_0 \approx 154.36$.

    (a) The coffee was about 154 degrees at time 0.

    (b) Room temperature is about 68 degrees.

13. (a) Since $\lim\limits_{t \to \infty} e^{-0.1t} = 0$, $\lim_{t \to \infty} y(t) = \lim_{t \to \infty}(T_0 - T_r)e^{-0.1t} + T_r = T_r$.

    (b) In coffee language, this means (as experience suggests!) that in the long run coffee cools to room temperature.

14. Solving algebraically for k gives $k = -\dfrac{\ln(\frac{19}{49})}{1000}$; an approximate decimal equivalent is 0.000947, as claimed.

15. Differentiating the right side of $P' = kP(C - P)$ with respect to P gives

    $$\frac{d}{dP}(kP(C - P)) = \frac{d}{dP}(kCP - kP^2) = kC - 2kP = 0 \iff P = \frac{C}{2}.$$

    Thus P' has its maximum value where $P = C/2$, as claimed. (In rumor language: The rumor spreads fastest when *half* the people know it.)

16. Let's solve $P(t) = 500$ for t (check the steps):

    $$P(t) = \frac{1000}{49e^{-0.947t} + 1} = 500 \implies 2 = 49e^{-0.947t} + 1 \implies t = \frac{\ln 49}{0.947} \approx 4.11.$$

    (In rumor language: After 4.11 days, half the people know the rumor.) (Notice that one could also have read this from the graph.)

17. $y' = y^2 \implies y'/y^2 = 1 \implies -1/y = x + C$ so $y = -1/(x + C)$. The initial condition $y(1) = 1$ implies that $C = -2$. Thus, the solution of the IVP is $y = -1/(x - 2)$.

18. $e^y = 2x + C$ so $y = \ln(2x + C)$. Thus, the solution of the IVP is $y = \ln(2x + e)$.

19. $y' = x^2 y \implies y'/y = x^2 \implies \ln y = x^3/3 + C$ so $y = Ae^{x^3/3}$. Thus, the solution of the IVP is $y = -2e^{x^3/3}$.

20. $y + y^3/3 = x^2/2 + C$ so the solution of the IVP is $y + y^3/3 = x^2/2 + 10$.

21. $y' - xy^2 = 0 \implies y'/y^2 = x \implies -1/y = x^2/2 + C$ so $y = 2/(A - x^2)$. Thus, the solution of the IVP is $y = 2/(3 - x^2)$.

22. $y^2/2 = x^2/2 + C$ or $y = \pm\sqrt{x^2 + A}$ so the solution of the IVP is $y = \sqrt{x^2 + 5}$.

23. (a) $f(y) = 1/y, g(x) = x$.

    (b) $F(y) = \ln y, G(x) = x^2/2$.

    (c) Differentiating both sides of the equation $F(y) = G(x) + C$ with respect to x leads to the equation $F'(y)y' = G'(x)$. Since $F' = f$ and $G' = g$, this equation is equivalent to the equation $f(y)y' = g(x)$. Thus, if y is implicitly defined by the equation $F(y) = G(x) + C$, y is a solution of the DE $y' = xy$.

    (d) Let $F(y) = G(x) + C$, where C is a constant. Then, using the results from parts (a)–(c), we have

$$\ln y = \frac{x^2}{2} + C \implies y = e^{x^2/2 + C} = Ke^{x^2/2}.$$

24. The point is that $e^{-x} \to 0$ as $x \to \infty$. Now if k and C are positive, $Ckt \to \infty$ as $t \to \infty$. Therefore $e^{-Ckt} \to 0$ as $t \to \infty$. Thus

$$\lim_{t \to \infty} P(t) = \lim_{t \to \infty} \frac{C}{Ke^{-Ckt} + 1} = \frac{C}{K \cdot 0 + 1} = C.$$

In biological terms, this means that in the long run, the population tends toward C, its carrying capacity.

25. (a) The exact values of k for curves P_1 to P_4 are, respectively, 2, 1.5, 1, and 0.5. (Thus the values of $-k$ are, respectively, -2, -1.5, -1, and -0.5.)

    The k-values can be found—approximately—by reading the graphs. The graph of P_1, for example, seems to pass through the point $(2, 750)$. This gives an equation we can solve for k, as follows:

$$P_1(2) = \frac{1000}{19e^{-2k} + 1} \approx 750 \implies k = \frac{\ln 57}{2} \approx 2.02.$$

    Values of k for the other curves are found similarly.

    (b) P_1 corresponds to the "hottest" rumor, P_4 to the "coolest." The graph shapes agree—hotter rumors spread faster.

26. $\arcsin y = x + C$ or $y = \sin(x + C)$.

27. $\arctan y = x + C$ or $y = \tan(x + C)$.

28. $-\ln|1 - y| = x + C$ or $y = 1 - Ae^{-x}$.

29. $\dfrac{1}{4} \ln \left| \dfrac{y + 2}{y - 2} \right| = x + C$ or $y = \dfrac{2Ae^{4x} + 2}{Ae^{4x} - 1}$.

30. $\ln(\ln y) = x + C$ or $y = e^{Ae^x}$.

31. $\ln\left(\sec y + \tan y\right) = x^3/3 + C$.

32.   (a)   From the problem statement, $\displaystyle\int_a^x \sqrt{1 + \left(f'(t)\right)^2}\, dt = \int_a^x f(t)\, dt$. Differentiating both

   sides of this equation with respect to x leads to the equation $\sqrt{1 + \left(f'(x)\right)^2} = f(x)$.

   (b)   Solving the equation from part (a) for $f'(x)$, we find that $y = f(x)$ must be a solution

   of the DE $y' = \sqrt{y^2 - 1}$. Since $\displaystyle\int \frac{dx}{\sqrt{x^2 - 1}} = \cosh^{-1} x$ and $\cosh 0 = 1$,

   $y(x) = \cosh x$ is a function with the desired properties.

§7.5 Present Value

1. For any interest rate r, the present value of one \$1 million payment, 23 years ahead, is $PV = 1,000,000 \cdot e^{-r \cdot 23}$.

   (a) If $r = 0.06$, then $PV = \$1,000,000 e^{-0.06 \cdot 23} \approx \$251,579$.

   (b) If $r = 0.08$, then $PV = \$1,000,000 e^{-0.08 \cdot 23} \approx \$158,817$.

   (c) To find the desired r we solve the equation $PV = 100,000 = 1,000,000 e^{-r \cdot 23}$ for r. The result: $r = (\ln 10)/23 \approx 0.10011$—just a bit above 10%.

2. For any interest rate r, the present value of one \$1 million payment, 45 years ahead, is $PV = 1,000,000 \cdot e^{-r \cdot 45}$. If $r = 0.06$, the formula gives $PV \approx 67,206$; if $r = 0.08$, $PV \approx 27,324$.

   If $r = 0.2$, then the formula gives $PV \approx 123.41$. (This preposterous result shows why 45-year bonds paying 20% are not (legally) available.)

3. For any interest rate r (real or nominal) the present value of one \$1 million payment, 23 years ahead, is $PV = 1,000,000 \cdot e^{-r \cdot 23}$.

   (a) If $r = 0.02$, then $PV = \$1,000,000 e^{-0.02 \cdot 23} \approx \$631,284$.

   (b) If $r = 0.04$, then $PV = \$1,000,000 e^{-0.04 \cdot 23} \approx \$398,519$.

   (c) To find Betty's r we solve the equation $PV = 200,000 = 1,000,000 e^{-r \cdot 23}$ for r. The result: $r = (\ln 5)/23 \approx 0.07$. Thus Betty needs to find a *real* interest rate—after inflation—of 7%.

4. For any interest rate r, the present value of one \$1 million payment, 45 years ahead, is $PV = 1,000,000 \cdot e^{-r \cdot 45}$. If $r = 0.02$, the formula gives $PV \approx 406,570$; if $r = 0.04$, $PV \approx 165,299$.

   If $r = 0.15$, then the formula gives $PV \approx 1,170.88$. (This preposterous result shows why 45-year investments paying 15% real interest are either extremely risky or illegal.)

5. At any interest rate r, the present value formula for several future payments says, in this situation, that

$$PV = 40,000 e^{-18r} + 42,000 e^{-19r} + 44,000 e^{-20r} + 46,000 e^{-21r}.$$

   If $r = 0.06$, the formula (and some electronic help) give $PV \approx \$53,316.85$. If $r = 0.08$, $PV \approx \$36,119.66$.

6. Let $t = 0$ denote 1993. Then the payment stream function $p(t) = 43,000$ applies for $18 \le t \le 22$ (i.e., from 2011 to 2015). For a given r the present value is the integral

$$PV = \int_{18}^{22} 43,000 e^{-rt}\, dt = -43,000 \frac{e^{-rt}}{r} \Big]_{18}^{22} = 43,000 \frac{e^{-18r} - e^{-22r}}{r}.$$

Substituting $r = 0.06$ gives $PV \approx \$51,929.83$. If $r = 0.08$, then $PV \approx \$53,316.85$. (For comparison: the four-lump-payment "discrete" model gave $PV \approx \$36,119.66$ if $r = 0.06$ and $PV \approx \$35,001.42$ if $r = 0.08$.)

7.   (a) p_I is the graph which peaks at $t = 180$. When $t = 180$, $p_I = 110$.

    (b) The function $\cos t$ has period 2π. The functions p_T and p_I both have period 360.

    (c) The constant 50 affects the amplitude of the graphs. The constant 60 shifts the graphs upward. The constants 180, 105, and π affect where each graph has its local maxima and minima.

8.   (a) $\displaystyle \int_{10}^{20} 12{,}000e^{-0.06t}\, dt \approx \$49{,}523.$

    (b) $\displaystyle \int_{10}^{20} 12{,}000e^{-0.08t}\, dt \approx \$37{,}115.$

    (c) $\displaystyle \int_{10}^{20} 12{,}000\, dt = \$120{,}000.$

    (d) $\displaystyle \int_{10}^{20} 12{,}000e^{-0.04t}\, dt \approx \$66{,}297.$

    (e) $\displaystyle \int_{10}^{20} 12{,}000e^{-0.06t}\, dt \approx \$49{,}523.$

9. The total return from the investment is \$40,000 paid continuously over an 8-year time interval. Since the present value of the return from the investment is
$PV = \$5{,}000 \int_0^8 e^{-0.06t}\, dt \approx \$31{,}768$, this is a worthwhile investment.

10. Using the midpoint rule with 50 subdivisions (and Maple): $PV_F \approx 21068.89$, $PV_T \approx 21203.64$, and $PV_I \approx 21067.78$.

11.   (a) $\displaystyle \left(\frac{e^{ax}}{a^2 + b^2} \big(a\cos(bx) + b\sin(bx)\big) \right)' = \frac{ae^{ax}}{a^2 + b^2} \big(a\cos(bx) + b\sin(bx)\big)$

$$+ \frac{e^{ax}}{a^2 + b^2}\big(-ab\sin(bx) + b^2\cos(bx)\big)$$

$$= e^{ax}\cos(bx).$$

    (b) $\displaystyle \int_0^{360} \left(50\cos\left(\pi \cdot \frac{t - 180}{180}\right) + 60 \right) e^{-0.1t/360}\, dt$

$$= \left(216{,}000 - \frac{180{,}000}{1 + 400\pi^2} \right) \big(1 - e^{-1/10}\big)$$

$$\approx 20{,}550.78.$$

$$\text{NOTE: } \cos\left(\pi \cdot \frac{t - 180}{180}\right) = \cos\left(\frac{\pi t}{180} - \pi\right) = -\cos\left(\frac{\pi t}{180}\right).$$

12. Yes. The present value of this investment stream is $\int \dfrac{1000}{1+t} e^{-0.05t}\, dt \approx 2065 > 2000$.

13. The interest earned from the income received at time t is $p(t)e^{r(T-t)}$. Thus, the total income accrued between $t = 0$ and $t = T$ is $\displaystyle\int_0^T p(t)e^{r(T-t)}\, dt = e^{rT}\int_0^T p(t)e^{-rt}\, dt = e^{rT}\, PV$.

8   *Symbolic Antidifferentiation Techniques*

§8.1   Integration by Parts

1. (a) Using the product rule, $(x \ln x)' = x \cdot 1/x + 1 \cdot \ln x = 1 + \ln x$.

   (b) Part (a) implies that $\int (1 + \ln x)\,dx = x + \int \ln x\,dx = x \ln x + C$. Thus,

   $$\int \ln x\,dx = x \ln x - x + C.$$

2. (a) $(x^4 \ln x)' = x^4 \cdot 1/x + 4x^3 \cdot \ln x = x^3 + 4x^3 \ln x$.

   (b) Part (a) implies that $\int (x^3 + 4x^3 \ln x)\,dx = x^4 \ln x + K$, so

   $$\int 4x^3 \ln x\,dx = x^4 \ln x - x^4/4 + K. \text{ Thus, } \int x^3 \ln x\,dx = \frac{x^4 \ln x}{4} - \frac{x^4}{16} + C.$$

   (c) If $u = \ln x$ and $dv = x^3\,dx$, then $du = dx/x$ and $v = x^4/4$. Thus, using integration by parts,

   $$\int x^3 \ln x\,dx = \ln x \cdot \frac{x^4}{4} - \int \frac{x^4}{4} \cdot \frac{dx}{x} = \frac{x^4 \ln x}{4} - \frac{1}{4} \int x^3\,dx = \frac{x^4 \ln x}{4} - \frac{x^4}{16} + C.$$

3. (a) Using the product and chain rules, $(xe^{-x})' = 1 \cdot e^{-x} + x \cdot (-1)e^{-x} = e^{-x} - xe^{-x}$.

   (b) Part (a) implies that $\int (e^{-x} - xe^{-x})\,dx = -e^{-x} - \int xe^{-x}\,dx = xe^{-x} + K$. Thus,

   $$\int xe^{-x}\,dx = -e^{-x} - xe^{-x} + C.$$

   (c) If $u = x$ and $dv = e^{-x}\,dx$, then $du = dx$ and $v = -e^{-x}$. Thus, using integration by parts,

   $$\int xe^{-x}\,dx = x \cdot (-e^{-x}) - \int (-e^{-x})\,dx = -xe^{-x} + \int e^{-x}\,dx = -xe^{-x} - e^{-x} + C.$$

4. (a) Using the product and chain rules,
   $(xe^{2x})' = 1 \cdot e^{2x} + x \cdot 2e^{2x} = e^{2x} + 2xe^{2x} = (1 + 2x)e^{2x}$.

   (b) Part (a) implies that $\int (1 + 2x)e^{2x}\,dx = \frac{e^{2x}}{2} + 2 \int xe^{2x}\,dx = xe^{2x} + K$. Thus,

   $$\int xe^{2x}\,dx = \frac{xe^{2x}}{2} - \frac{e^{2x}}{4} + C.$$

   (c) If $u = x$ and $dv = e^{2x}\,dx$, then $du = dx$ and $v = e^{2x}/2$. Thus, using integration by parts,

   $$\int xe^{2x}\,dx = x \cdot \frac{e^{2x}}{2} - \frac{1}{2} \int e^{2x}\,dx = \frac{xe^{2x}}{2} - \frac{e^{2x}}{4} + C.$$

5.   (a) Using the product and chain rules,

$$\left(x \arctan(2x)\right)' = 1 \cdot \arctan(2x) + x \cdot \frac{2}{1 + (2x)^2} = \arctan(2x) + \frac{2x}{1 + 4x^2}.$$

(b) Part (a) implies that

$$\int \left(\arctan(2x) + \frac{2x}{1 + 4x^2}\right) dx = \int \arctan(2x)\, dx + \frac{1}{4} \ln(1 + 4x^2)$$

$$= x \arctan(2x) + C.$$

Thus, $\int \arctan(2x)\, dx = x \arctan(2x) - \frac{1}{4} \ln(1 + 4x^2) + C.$

(c) If $u = \arctan(2x)$ and $dv = dx$, then $du = (2\, dx)/(1 + 4x^2)$ and $v = x$. Thus, using integration by parts,

$$\int \arctan(2x)\, dx = x \arctan(2x) - \int \frac{2x\, dx}{1 + 4x^2} = x \arctan(2x) - \frac{1}{4} \ln(1 + 4x^2) + C.$$

6.   (a) Using the product rule,

$$(x \arcsin x)' = 1 \cdot \arcsin x + x \cdot \frac{1}{\sqrt{1 - x^2}} = \arcsin x + \frac{x}{\sqrt{1 - x^2}}.$$

(b) Part (a) implies that

$$\int \left(\arcsin x + \frac{x}{\sqrt{1 - x^2}}\right) dx = \int \arcsin x\, dx - \sqrt{1 - x^2} = x \arcsin x + C \text{ so}$$

$$\int \arcsin x\, dx = x \arcsin x + \sqrt{1 - x^2} + C.$$

(c) If $u = \arcsin x$ and $dv = dx$, then $du = dx/\sqrt{1 - x^2}$ and $v = x$. Thus, using integration by parts,

$$\int \arcsin x\, dx = x \arcsin x - \int \frac{x}{\sqrt{1 - x^2}}\, dx = x \arcsin x + \sqrt{1 - x^2} + C.$$

7.   (a) Using the product and chain rules,
     $(e^{2x} \sin x)' = 2e^{2x} \cdot \sin x + e^{2x} \cdot \cos x = (2 \sin x + \cos x)e^{2x}.$

(b) Using the product and chain rules,
     $(e^{2x} \cos x)' = 2e^{2x} \cdot \cos x + e^{2x} \cdot (-\sin x) = (2 \cos x - \sin x)e^{2x}.$

(c) Part (a) implies that

$$\int (2 \sin x + \cos x)e^{2x}\, dx = 2 \int e^{2x} \sin x\, dx + \int e^{2x} \cos x\, dx = e^{2x} \sin x + K, \text{ so}$$

$$2 \int e^{2x} \sin x\, dx = e^{2x} \sin x - \int e^{2x} \cos x\, dx + K. \text{ Now, part (b) implies that}$$

$\int e^{2x} \sin x \, dx = 2 \int e^{2x} \cos x \, dx - e^{2x} \cos x$. Combining these results,

$$2 \cdot 2 \int e^{2x} \sin x \, dx + \int e^{2x} \sin x \, dx = 2e^{2x} \sin x - 2 \int e^{2x} \cos x \, dx + 2K$$

$$+ 2 \int e^{2x} \cos x \, dx - e^{2x} \cos x$$

$$= e^{2x}(2 \sin x - \cos x) + 2K.$$

Thus, $\int e^{2x} \sin x \, dx = \dfrac{e^{2x}}{5}(2 \sin x - \cos x) + C$.

(d) Let $u = \sin x$ and $dv = e^{2x} \, dx$. Then, $du = \cos x \, dx$ and $v = e^{2x}/2$ so

$$\int e^{2x} \sin x \, dx = \frac{e^{2x}}{2} \sin x - \frac{1}{2} \int e^{2x} \cos x \, dx.$$

Now, if $u = \cos x$ and $dv = e^{2x} \, dx$, $du = -\sin x \, dx$ and $v = e^{2x}/2$ so

$$\int e^{2x} \cos x \, dx = \frac{e^{2x}}{2} \cos x + \frac{1}{2} \int e^{2x} \sin x \, dx.$$

Thus, $\displaystyle \int e^{2x} \sin x \, dx = \frac{e^{2x}}{2} \sin x - \frac{1}{2} \left(\frac{e^{2x}}{2} \cos x + \frac{1}{2} \int e^{2x} \sin x \, dx \right)$

$$\implies \frac{5}{4} \int e^{2x} \sin x \, dx = \frac{e^{2x}}{4}(2 \sin x - \cos x) + C$$

$$\implies \int e^{2x} \sin x \, dx = \frac{e^{2x}}{5}(2 \sin x - \cos x) + C.$$

8. (a) Using the product and chain rules,

$$\left(e^x \cos(2x) \right)' = e^x \cdot \cos(2x) + e^x \cdot 2(-\sin(2x)) = e^x(\cos(2x) - 2\sin(2x)).$$

(b) Using the product and chain rules,

$$\left(e^x \sin(2x) \right)' = e^x \cdot \sin(2x) + e^x \cdot 2\cos(2x) = e^x(\sin(2x) + 2\cos(2x)).$$

(c) Part (a) implies that $\displaystyle \int e^x \cos(2x) \, dx = e^x \cos(2x) + 2 \int e^x \sin(2x) \, dx$. Part (b)

implies that $\displaystyle \int e^x \sin(2x) \, dx = e^x \sin(2x) - 2 \int e^x \cos(2x) \, dx$. Thus,

$\displaystyle \int e^x \cos(2x) \, dx = e^x \cos(2x) + 2e^x \sin(2x) - 4 \int e^x \cos(x) \, dx$ so

$\displaystyle \int e^x \cos(2x) \, dx = e^x(\cos(2x) + 2\sin(2x))/5.$

(d) Using integration by parts (twice):

$$\int e^x \cos(2x)\,dx = e^x \cos(2x) + 2\int e^x \sin(2x)\,dx$$

$$= e^x \cos(2x) + 2e^x \sin(2x) - 4\int e^x \cos(2x)$$

Thus, $5\int e^x \cos(2x)\,dx = e^x \cos(2x) + 2e^x \sin(2x)$ so

$$\int e^x \cos(2x)\,dx = e^x\big(\cos(2x) + 2\sin(2x)\big)/5 + C.$$

9. $\displaystyle\int x \sin(3x)\,dx = \frac{\sin(3x)}{9} - \frac{x\cos(3x)}{3} + C$ $[du = dx, v = -\frac{1}{3}\cos(3x)]$.

10. $\displaystyle\int \frac{\ln x}{x^2}\,dx = -\frac{\ln x}{x} + \int \frac{dx}{x^2} = -\frac{\ln x}{x} - \frac{1}{x} + C = -\frac{1 + \ln x}{x} + C$

     $[du = dx/x, \; v = -1/x]$.

11. $\displaystyle\int x \sec^2 x\,dx = x \tan x + \ln|\cos x| + C$ $[du = dx, v = \tan x]$.

12. $\displaystyle\int x \sec x \tan x\,dx = x \sec x - \int \sec x\,dx = x \sec x - \ln|\sec x + \tan x| + C$

     $[du = dx, \; v = \sec x]$.

13. $\displaystyle\int (\ln x)^2\,dx = (\ln x)(x \ln x - x) - \int (x \ln x - x)\frac{dx}{x}$

$$= (\ln x)(x \ln x - x) - (x \ln x - x) + x + C$$

$$= x(\ln x)^2 - 2x \ln x + 2x + C \quad [du = dx/x, v = x \ln x - x].$$

14. Using integration by parts with $du = dx/(2\sqrt{x})$ and $v = x \ln x - x$

$$\int \sqrt{x} \ln x\,dx = \sqrt{x}(x \ln x - x) - \int (x \ln x - x)\frac{dx}{2\sqrt{x}}$$

$$= \sqrt{x}(x \ln x - x) - \frac{1}{2}\int \left(\sqrt{x} \ln x - \sqrt{x}\right)dx$$

$$= \sqrt{x}(x \ln x - x) - \frac{1}{2}\int \sqrt{x} \ln x\,dx + \frac{x^{3/2}}{3}$$

$$= x^{3/2} \ln x - \frac{1}{2}\int \sqrt{x} \ln x\,dx - \frac{2x^{3/2}}{3}.$$

Thus, $\displaystyle\int \sqrt{x} \ln x\,dx = \frac{2x^{3/2} \ln x}{3} - \frac{4x^{3/2}}{9} + C.$

15. $\int_0^1 xe^{-x}\,dx = -(1+x)e^{-x}\Big]_0^1 = 1 - 2e^{-1} \approx 0.26424.\quad [M_{20} \approx 0.26435].$

16. $\int_0^\pi x\cos(2x)\,dx = \frac{1}{2}x\sin(2x) + \frac{1}{4}\cos(2x)\Big]_0^\pi = 0.\quad [M_{20} = 0].$

17. $\int_1^e x\ln x\,dx = \frac{1}{2}x^2\ln x - \frac{1}{4}x^2\Big]_1^e = \frac{1}{4}\left(e^2+1\right) \approx 2.09726.\quad [M_{20} \approx 2.0970].$

18. $\int_{\pi/4}^{\pi/2} x\csc^2 x\,dx = -x\cot x + \ln(\sin x)\Big]_{\pi/4}^{\pi/2} = \pi/4 - \ln\left(\sqrt{2}/2\right)$

$$= \pi/4 + \ln\left(\sqrt{2}\right) \approx 1.13197.\ [M_{20} \approx 1.1318].$$

19. $\int_{-1}^{\sqrt{2}/2} x^2\arctan x\,dx = \frac{1}{3}x^3\arctan x - \frac{1}{6}x^2 + \frac{1}{6}\ln\left(1+x^2\right)\Big]_{-1}^{\sqrt{2}/2}$

$$= \frac{\sqrt{2}}{12}\arctan(\sqrt{2}/2) + \frac{1}{12} + \frac{1}{6}\ln(3/4) - \frac{\pi}{12} \approx -0.15388.\quad [M_{20} \approx -0.15361].$$

20. $\int_1^4 e^{3x}\cos(2x)\,dx = \frac{3}{13}e^{3x}\cos(2x) + \frac{2}{13}e^{3x}\sin(2x)\Big]_1^4 \approx 19{,}307.\quad [M_{20} \approx 19{,}671].$

21. (a) If $u = x^2$ and $dv = \cos x\,dx$, then $du = 2x\,dx$ and $v = \sin x$. Thus, using integration by parts, $\int x^2\cos x\,dx = x^2\sin x - 2\int x\sin x\,dx.$

    (b) Let $u = x$ and $dv = \sin x\,dx$. Then, $du = dx$ and $v = -\cos x$ so

$$\int x\sin x\,dx = -x\cos x + \int \cos x\,dx = -x\cos x + \sin x + C.$$

    (c) Combining parts (a) and (b),

$$\int x^2\cos x\,dx = x^2\sin x + 2x\cos x - 2\sin x + C.$$

22. (a) If $u = x^2$ and $dv = \sin x\,dx$, then $du = 2x\,dx$ and $v = -\cos x$. Thus, using integration by parts, $\int x^2\sin x\,dx = -x^2\cos x + 2\int x\cos x\,dx.$

    (b) $\int x\cos x\,dx = x\sin x + \int \sin x\,dx = x\sin x + \cos x + C\quad [u = x, dv = \cos x\,dx].$

    (c) Combining the results in parts (a) and (b),

$$\int x^2\sin x\,dx = -x^2\cos x + 2x\sin x + 2\cos x + C.$$

23. Since $n = 3$, repeated use of the reduction formula yields

$$\int (\ln x)^3 \, dx = x \, (\ln x)^3 - 3 \int (\ln x)^2 \, dx$$

$$= x \, (\ln x)^3 - 3x \, (\ln x)^2 + 6 \int \ln x \, dx$$

$$= x \, (\ln x)^3 - 3x \, (\ln x)^2 + 6x(\ln x - 1) + C.$$

24. $\displaystyle \int x \, (\ln x)^2 \, dx = \frac{1}{2} x^2 \, (\ln x)^2 - \int x \ln x \, dx$

$$= \frac{1}{2} x^2 \, (\ln x)^2 - \frac{1}{2} x^2 \ln x + \frac{x^2}{4} + C.$$

25. Using integration by parts with $u = \ln x$, $dv = x^r \, dx$, $du = dx/x$ and $v = x^{r+1}/(r+1)$:

$$\int x^r \ln x \, dx = \frac{x^{r+1} \ln x}{r+1} - \frac{1}{r+1} \int x^{r+1} \frac{dx}{x} = \frac{x^{r+1} \ln x}{r+1} - \frac{1}{r+1} \int x^r \, dx$$

$$= \frac{x^{r+1}}{r+1} \left(\ln x - \frac{1}{r+1} \right) + C.$$

26. Let $u = x^n$ and $dv = e^x \, dx$. Then, $du = nx^{n-1} \, dx$ and $v = e^x$. Therefore,

$$\int x^n e^x \, dx = x^n e^x - n \int x^{n-1} e^x \, dx.$$

27. Let $u = (\ln x)^n$ and $dv = dx$. Then $du = n \, (\ln x)^{n-1} \, dx$ and $v = x$. Therefore,

$$\int (\ln x)^n \, dx = x \, (\ln x)^n - n \int (\ln x)^{n-1} \, dx, \quad n \geq 1.$$

28. Let $u = (\ln x)^n$, $dv = x \, dx$, $du = n(\ln x)^{n-1} \cdot \dfrac{1}{x} \, dx$, and $v = \dfrac{x^2}{2}$. Then integration by parts implies that

$$\int x \, (\ln x)^n \, dx = \frac{(\ln x)^n x^2}{2} - \frac{n}{2} \int x(\ln x)^{n-1} \, dx, \quad n \geq 1.$$

29. (a) If $u = x^3$, then $du = 3x^2 \, dx$, so
$$\int x^2 \ln(x^3) \, dx = \frac{1}{3} \int \ln u \, du = \frac{u}{3}(\ln u - 1) = \frac{x^3}{3}\left(\ln(x^3) - 1\right).$$

   (b) Let $u = \ln x$, $dv = x^2 \, dx$, $du = dx/x$, and $v = x^3/3$. Then, using integration by parts,

$$\int x^2 \ln(x^3) \, dx = 3 \int x^2 \ln x \, dx = x^3 \ln x - \int x^2 \, dx = x^3 \ln x - \frac{x^3}{3}.$$

30.   (a) If $u = a + bx$, then $du = b\,dx$. Therefore,

$$\int x\sqrt{a+bx}\,dx = \frac{1}{b^2}\int (u-a)\sqrt{u}\,du = \frac{2}{b^2}\left(\frac{u^{5/2}}{5} - \frac{au^{3/2}}{3}\right)$$

$$= \frac{2}{b^2}\left(\frac{(a+bx)^{5/2}}{5} - \frac{a(a+bx)^{3/2}}{3}\right) + C.$$

(b) If $u = x$ and $dv = \sqrt{a+bx}\,dx$, then $du = dx$ and $v = 2(a+bx)^{3/2}/3b$. Therefore, using integration by parts,

$$\int x\sqrt{a+bx}\,dx = \frac{2x(a+bx)^{3/2}}{3b} - \frac{2}{3b}\int (a+bx)^{3/2}\,dx$$

$$= \frac{2x(a+bx)^{3/2}}{3b} - \frac{4(a+bx)^{5/2}}{15b^2} + C.$$

31. $\displaystyle\int x^3 e^{x^2}\,dx = \int x\cdot x^2 e^{x^2}\,dx = \frac{1}{2}\int u e^u\,du$

$$= \frac{1}{2}\left(ue^u - \int e^u\,du\right) = (u-1)e^u/2 = (x^2-1)e^{x^2}/2.$$

32. $\displaystyle\int x^5 e^{-x^3}\,dx = \int x^2\cdot x^3 e^{-x^3}\,dx = -\frac{1}{3}\int u e^u\,du$

$$= \frac{1}{3}\left(ue^u - e^u\right) = \frac{(u-1)e^u}{3}$$

$$= \frac{(-x^3-1)}{3}e^{-x^3} = -\frac{(x^3+1)e^{-x^3}}{3} + C.$$

33.   (a) Let $u = \sin x$ and $dv = \sin x\,dx$. Then, $du = \cos x\,dx$ and $v = -\cos x$ so
$$\int \sin^2 x\,dx = -\sin x\cos x + \int \cos^2 x\,dx.$$

(b) $\displaystyle\int \sin^2 x\,dx = -\sin x\cos x + \int \cos^2 x\,dx = -\sin x\cos x + \int (1-\sin^2 x)\,dx.$

Therefore, $2\displaystyle\int \sin^2 x\,dx = x - \sin x\cos x$ and $\displaystyle\int \sin^2 x\,dx = \frac{1}{2}(x - \sin x\cos x) + C.$

34. Let $u = \cos x$ and $dv = \cos x\,dx$. Then, $du = -\sin x\,dx$ and $v = \sin x$ so

$$\int \cos^2 x\,dx = \int \cos x\cos x\,dx = \cos x\sin x + \int \sin^2 x\,dx$$

$$= \cos x\sin x + \int (1-\cos^2 x)\,dx$$

$$= \cos x\sin x + x - \int \cos^2 x\,dx.$$

Thus, $2 \int \cos^2 x \, dx = \cos x \sin x + x$ which implies that

$$\int \cos^2 x \, dx = (\cos x \sin x + x)/2 + C.$$

35. Let $u = x$ and $dv = \cos^2 x \, dx = \frac{1}{2}\left(1 + \cos(2x)\right) dx$. Then $du = dx$ and $v = \frac{1}{2}\left(x + \frac{1}{2}\sin(2x)\right)$. Therefore,

$$\int x \cos^2 x \, dx = \frac{1}{2}x^2 + \frac{1}{4}x \sin(2x) - \frac{1}{2}\int \left(x + \frac{\sin(2x)}{2}\right) dx$$

$$= \frac{1}{4}x^2 + \frac{1}{4}x \sin(2x) + \frac{1}{8}\cos(2x) + C.$$

36. Using the identity given in the hint,

$$\int x \sin^2 x \, dx = \frac{1}{2}\int x\left(1 - \cos(2x)\right) dx = x^2/4 - \frac{1}{2}\int x \cos(2x)\, dx.$$ The latter antiderivative can be found using integration by parts (with $u = x$ and $dv = \cos(2x)\,dx$):

$$\int x \cos(2x)\, dx = \frac{x \sin(2x)}{2} - \frac{1}{2}\int \sin(2x)\, dx = \frac{x \sin(2x)}{2} + \frac{\cos(2x)}{4} + C.$$

Therefore, $\int x \sin^2 x \, dx = \frac{1}{4}\left(x^2 - x\sin(2x) - \frac{1}{2}\cos(2x)\right) + C.$

37. $\int x^2 \cos x \, dx = x^2 \sin x - 2\sin x + 2x \cos x + C.$

Let $u = x^2$, $dv = \cos x \, dx$. Then, $\int x^2 \cos x \, dx = x^2 \sin x - 2\int x \sin x \, dx.$ The remaining antiderivative can be found using integration by parts again.

38. $\int x \sin x \cos x \, dx = \frac{1}{4}\cos x \sin x - \frac{1}{2}x \cos^2 x + \frac{1}{4}x + C$

$[u = x, \, dv = \sin x \cos x \, dx, \, v = \frac{1}{2}\cos^2 x].$

39. Let $u = x$ and $dv = \csc x \cot x \, dx$. Then $du = dx$ and $v = -\csc x$ so

$$\int x \csc x \cot x \, dx = -x \csc x + \int \csc x \, dx = -x \csc x + \ln|\csc x - \cot x| + C.$$

40. $\int \arccos x \, dx = x \arccos x - \sqrt{1 - x^2} + C \quad [u = \arccos x, \, dv = dx].$

41. $\int \sqrt{x} \ln\left(\sqrt[3]{x}\right) dx = \frac{1}{3}\int \sqrt{x} \ln x \, dx = \frac{2}{9}x^{3/2}\ln x - \frac{4}{27}x^{3/2} + C.$

$[u = \ln x, \, dv = \sqrt{x}\,dx].$

42. $\int x e^x \sin x \, dx = \frac{1}{2}(1 - x)e^x \cos x + \frac{1}{2}x e^x \sin x + C. \quad [u = x, \, dv = e^x \sin x \, dx].$

43. $\int \arctan(1/x)\, dx = x\arctan(1/x) + \dfrac{1}{2}\ln\left(x^2 + 1\right) + C.$ $[u = \arctan(1/x), dv = dx]$.

44. First, make the substitution $u = \ln x, du = dx/x$ so $\int x^3(\ln x)^2\, dx = \int u^2 e^{4u}\, du$. The latter antiderivative can be found by repeated use of integration by parts:

$$\int u^2 e^{4u}\, du = u^2 e^{4u}/4 - \frac{1}{2}\int u e^{4u}\, du$$

$$= u^2 e^{4u}/4 - u e^{4u}/8 + \frac{1}{8}\int e^{4u}\, du = u^2 e^{4u}/4 - u e^{4u}/8 + e^{4u}/32 + C$$

$$= \frac{x^4}{4}\left((\ln x)^2 - \frac{\ln x}{2} + \frac{1}{8}\right) + C.$$

45. $\int e^{\sqrt{x}}\, dx = 2e^{\sqrt{x}}\sqrt{x} - 2e^{\sqrt{x}} + C.$
First substitute $w = \sqrt{x}, w^2 = x, 2w\, dw = dx$. This produces the new integral $\int 2w e^w\, dw$. Now use parts, with $u = w, dv = e^w\, dw$. The answer follows directly.

46. $\int x^5 \sin\left(x^3\right)\, dx = \dfrac{\sin\left(x^3\right)}{3} - \dfrac{x^3 \cos\left(x^3\right)}{3} + C.$
First substitute $w = x^3, dw = 3x^2\, dx$, to get the new integral $\frac{1}{3}\int w\sin w\, dw$. Now use parts, with $u = w, dv = \sin w\, dw$.

47. $\int \sin(\ln x)\, dx = \dfrac{x}{2}\big(\sin(\ln x) - \cos(\ln x)\big) + C.$
First substitute $w = \ln x, x = e^w, dx = e^w\, dw$. This gives the new integral
$\int \sin(\ln x)\, dx = \int e^w \sin w\, dw$. The new integral is done (with integration by parts, twice) in the book.

48. $\int \sqrt{x}e^{-\sqrt{x}}\, dx = -2e^{-\sqrt{x}}\left(x + 2\sqrt{x} + 2\right) + C.$

First substitute $w = \sqrt{x}, w^2 = x, 2w\, dw = dx$. This gives the new integral $2\int w^2 e^{-w}\, dw$. Finding the latter integral requires using integrations by parts twice.

49. $\int \sin\left(\sqrt{x}\right)\, dx = 2\sin\left(\sqrt{x}\right) - 2\sqrt{x}\cos\left(\sqrt{x}\right) + C.$
First substitute $w = \sqrt{x}$, then use integration by parts with $u = w$ and $dv = \sin w\, dw$.

50. $\int \dfrac{\arctan\left(\sqrt{x}\right)}{\sqrt{x}}\, dx = 2\sqrt{x}\arctan\left(\sqrt{x}\right) - \ln(1 + x) + C.$
Substitute $w = \sqrt{x}$, then use integration by parts with $u = \arctan w$ and $dv = dw$.

51. $\int \sqrt{x}\arctan\left(\sqrt{x}\right)\, dx = \dfrac{2}{3}x^{3/2}\arctan\left(\sqrt{x}\right) - \dfrac{1}{3}(1 + x) + \dfrac{1}{3}\ln(1 + x) + C.$
Substitute $w = \sqrt{x}$, then use integration by parts with $u = \arctan w$ and $dv = w^2\, dw$.

52. Using integration by parts with $u = \arcsin x$ and $dv = x^2\,dx$,

$$\int x^2 \arcsin x\,dx = \frac{1}{3}x^3 \arcsin x - \frac{1}{3}\int \frac{x^3}{\sqrt{1-x^2}}\,dx.$$

To find the remaining antiderivative, use the substitution $w = 1 - x^2$:

$$\int \frac{x^3}{\sqrt{1-x^2}}\,dx = -\frac{1}{2}\int \frac{1-w}{\sqrt{w}}\,dw = -\frac{1}{2}\int \frac{dw}{\sqrt{w}} + \frac{1}{2}\int \sqrt{w}\,dw$$

$$= -\sqrt{w} + \frac{1}{3}w^{3/2} + C = -\sqrt{1-x^2} + \frac{1}{3}(1+x^2)^{3/2} + C.$$

Therefore, $\displaystyle\int x^2 \arcsin x\,dx = \frac{1}{3}x^3 \arcsin x + \frac{1}{3}\sqrt{1-x^2} - \frac{1}{9}(1+x^2)^{3/2} + C.$

53. First, make the substitution $u = \sin x$. Then,

$$\int \cos x \ln(\sin x)\,dx = \int \ln u\,du = u\ln u - u = (\sin x)\ln(\sin x) - \sin x + C.$$

54. First, make the substitution $u = -x^2$, then use integration by parts:

$$\int x^5 e^{-x^2}\,dx = \int x \cdot x^4 e^{-x^2}\,dx = -\frac{1}{2}\int u^2 e^u\,du$$

$$= -\frac{1}{2}\left(u^2 e^u - 2\int u e^u\,du\right) = -\frac{1}{2}\left(u^2 e^u - 2u e^u + 2e^u\right) + C$$

$$= -\left(x^4/2 + x^2 + 1\right)e^{-x^2} + C.$$

55. Using integration by parts with $u = x$ and $dv = \sinh x\,dx$,

$$\int x \sinh x\,dx = x \cosh x - \int \cosh x\,dx = x \cosh x - \sinh x + C.$$

56. Using integration by parts with $u = x$ and $dv = \cosh(2x)\,dx$,

$$\int x \cosh(2x)\,dx = \frac{1}{2}\left(x \sinh(2x) - \int \sinh(2x)\,dx\right)$$

$$= \frac{1}{2}\left(x \sinh(2x) - \frac{1}{2}\cosh(2x)\right) + C.$$

57. Using integration by parts (first with $u = x^2$ and $dv = \sinh x\,dx$, then with $u = x$ and $dv = \cosh x$),

$$\int x^2 \sinh x\,dx = x^2 \cosh x - 2\int x \cosh x\,dx = x^2 \cosh x - 2x \sinh x + 2 \cosh x + C.$$

58. Make the substitution $u = \sqrt{x}$, then use integration by parts:

$$\int \cosh(\sqrt{x})\,dx = 2\int u \cosh u\,du = 2u \sinh u - 2\int \sinh u = 2u \sinh u - 2 \cosh u$$

$$= 2\sqrt{x}\sinh(\sqrt{x}) - 2\cosh(\sqrt{x}) + C.$$

59.   (a) $I_1 = \int_1^e \ln x \, dx = \left(x \ln x - x \right)\Big|_1^e = (e - e) - (0 - 1) = 1.$

(b) Using the reduction formula,

$$I_2 = \int_1^e (\ln x)^2 \, dx = x \, (\ln x)^2 \Big|_1^e - 2 \int_1^e \ln x \, dx = (e - 0) - 2I_1 = e - 2.$$

(c) Using the reduction formula,

$$I_3 = \int_1^e (\ln x)^3 \, dx = x(\ln x)^3 \Big|_1^e - 3 \int_1^e (\ln x)^2 \, dx = (e - 0) - 3I_2 = 6 - 2e.$$

(d) The reduction formula implies that $I_n = e - nI_{n-1}$. Thus, $I_4 = e - 4I_3 = 9e - 24$ and $I_5 = e - 5I_4 = 120 - 44e.$

60.   (a) $I_0 = \int_0^1 e^x \, dx = e^x \Big|_0^1 = e^1 - e^0 = e - 1.$

(b) Using the reduction formula, $I_1 = \int_0^1 x e^x \, dx = x e^x \Big|_0^1 - I_0 = e - (e - 1) = 1.$

(c) Using the reduction formula, $I_2 = \int_0^1 x^2 e^x \, dx = x^2 e^x \Big|_0^1 - 2I_1 = e - 2.$

(d) The reduction formula implies that $I_n = e - nI_{n-1}$. Thus, $I_3 = e - 3I_2 = 6 - 2e,$ $I_4 = e - 4I_3 = 9e - 24,$ and $I_5 = e - 5I_4 = 120 - 44e.$

61. $f(x) = x^4/4.$ Integration by parts with $u = \cos x$ and $dv = x^3 \, dx$ leads to $\int x^3 \cos x \, dx = \frac{1}{4} x^4 \cos x + \int \frac{1}{4} x^4 \sin x \, dx.$

62. An integration by parts with $u = f(x)$ and $dv = \sin x \, dx$ shows that

$$\int_0^\pi f(x) \sin x \, dx = -f(x) \cos x \Big]_0^\pi + \int_0^\pi f'(x) \cos x \, dx = f(\pi) + f(0) + \int_0^\pi f'(x) \cos x \, dx.$$

An integration by parts with $u = \sin x$ and $dv = f''(x) \, dx$ shows that

$$\int_0^\pi f''(x) \sin x \, dx = f'(x) \sin x \Big]_0^\pi - \int_0^\pi f'(x) \cos x \, dx = - \int_0^\pi f'(x) \cos x \, dx.$$

Combining these results, we have

$$6 = \int_0^\pi f(x) \sin x \, dx + \int_0^\pi f''(x) \sin x \, dx = f(\pi) + f(0) = f(\pi) + 2.$$

From this it follows that $f(\pi) = 4.$

63. Using integration by parts (twice), we can show that

$$\int f(x) \cos x \, dx = f(x) \sin x - \int f'(x) \sin x \, dx$$

$$= f(x) \sin x + f'(x) \cos x - \int f''(x) \cos x \, dx.$$

Thus,

$$\int_{-\pi/2}^{3\pi/2} f(x) \cos x \, dx = \left(f(x) \sin x + f'(x) \cos x \right) \Big]_{-\pi/2}^{3\pi/2} - \int_{-\pi/2}^{3\pi/2} f''(x) \cos x \, dx$$

$$= -f(3\pi/2) + f(-\pi/2) - 4 = -\int_{-\pi/2}^{3\pi/2} f'(x) \, dx - 4 = -5.$$

64. Suppose that v_1 and v_2 are both antiderivatives of dv. Then, $\int u \, dv = uv_1 - \int v_1 \, du$. Since v_2 is also an antiderivative of dv, $v_2 = v_1 + C$, where C is a constant. Therefore,

$$uv_2 - \int v_2 \, du = u(v_1 + C) - \int (v_1 + C) \, du$$

$$= uv_1 + Cu - \int v_1 \, du - Cu = uv_1 - \int v_1 \, du = \int u \, dv.$$

65. (a) This is the fundamental theorem of calculus.

  (b) Using integration by parts with $u = f'(x)$, $dv = dx$, and $v = x - b$, leads to

$$\int_a^b f'(x) \, dx = (x - b) f'(x) \Big|_a^b - \int_a^b (x - b) f''(x) \, dx$$

$$= (b - a) f'(a) - \int_a^b (x - b) f''(x) \, dx.$$

  Combining this result with part (a) implies that

$$f(b) = f(a) + f'(a)(b - a) - \int_a^b (x - b) f''(x) \, dx.$$

  (c) Using integration by parts with $u = f''(x)$, $dv = (x - b)dx$, and $v = (x - b)^2/2$, leads to

$$\int_a^b (x - b) f''(x) \, dx = \frac{1}{2}(x - b)^2 f''(x) \Big|_a^b - \frac{1}{2} \int_a^b (x - b)^2 f'''(x) \, dx$$

$$= -\frac{1}{2}(b - a)^2 f''(a) - \frac{1}{2} \int_a^b (x - b)^2 f'''(x) \, dx.$$

  Combining this with part (b) implies that

$$f(b) = f(a) + f'(a)(b - a) + \frac{1}{2} f''(a)(b - a)^2 + \frac{1}{2} \int_a^b (x - b)^2 f'''(x) \, dx.$$

(d) Using integration by parts with $u = f'''(x)$, $dv = (x - b)dx$, and $v = (x - b)^3/3$, leads to

$$\int_a^b (x - b)^2 f'''(x)\, dx = \frac{1}{3}(x - b)^3 f'''(x)\Big|_a^b - \frac{1}{3}\int_a^b (x - b)^3 f^{(4)}(x)\, dx$$

$$= \frac{1}{3}f'''(a)(b - a)^3 - \frac{1}{3}\int_a^b (x - b)^3 f^{(4)}(x)\, dx.$$

Thus,

$$f(b) = f(a) + f'(a)(b - a) + \frac{1}{2}f''(a)(b - a)^2 + \frac{1}{6}f'''(a)(b - a)^3$$

$$- \frac{1}{6}\int_a^b (x - b)^3 f^{(4)}(x)\, dx$$

$$= p_3(b) - \frac{1}{6}\int_a^b (x - b)^3 f^{(4)}(x)\, dx$$

so $f(b) - p_3(b) = -\dfrac{1}{6}\displaystyle\int_a^b (x - b)^3 f^{(4)}(x)\, dx.$

§8.2 Partial Fractions

1. (a) $\dfrac{1}{1+x} + \dfrac{1}{1-x} = \dfrac{(1-x)+(1+x)}{(1+x)(1-x)} = \dfrac{2}{1-x^2}.$

    (b) $\displaystyle\int \dfrac{dx}{1-x^2} = \dfrac{1}{2}\int \left(\dfrac{1}{1+x} + \dfrac{1}{1-x}\right) dx$

    $\qquad\qquad = \dfrac{\ln|1+x|}{2} - \dfrac{\ln|1-x|}{2} + C = \dfrac{1}{2}\ln\left|\dfrac{1+x}{1-x}\right| + C.$

2. $\dfrac{1}{x} + \dfrac{1}{1+x^2} + \dfrac{2}{x+2} = \dfrac{(1+x^2)(x+2) + x(x+2) + 2x(1+x^2)}{x(1+x^2)(x+2)}$

    $\qquad = \dfrac{(2+x++2x^2+x^3) + (x^2+2x) + (2x+2x^3)}{x(1+x^2)(x+2)} = \dfrac{2+5x+3x^2+3x^3}{x(1+x^2)(x+2)}.$

3. $\dfrac{x^3}{1+x^2} = \dfrac{x(1+x^2)-x}{1+x^2} = x - \dfrac{x}{1+x^2}.$

4. (a) $\dfrac{x}{1+x} = \dfrac{(x+1)-1}{x+1} = 1 - \dfrac{1}{x+1}.$

    (b) $\dfrac{x^2}{1+x} = \dfrac{x(1+x)-x}{1+x} = x - \dfrac{x}{1+x} = x - 1 + \dfrac{1}{1+x}.$

5. (a) $\dfrac{2}{x+1} + \dfrac{3}{x+2} = \dfrac{2(x+2)+3(x+1)}{(x+1)(x+2)} = \dfrac{5x+7}{(x+1)(x+2)}.$

    (b) $\displaystyle\int \dfrac{5x+7}{(x+1)(x+2)}\,dx = 2\ln|x+1| + 3\ln|x+2| + C.$

6. (a) No — combining the terms of the expression $\dfrac{A}{x} + \dfrac{B}{x+1}$ yields a rational expression with $x(x+1)$ in the denominator.

    (b) No. $\dfrac{A}{x^2} + \dfrac{B}{x+1} = \dfrac{Bx^2+Ax+A}{x^2(x+1)}$ so no choice of A and B will reproduce the numerator of p.

    (c) $\dfrac{A}{x} + \dfrac{B}{x^2} + \dfrac{C}{x+1} = \dfrac{(A+C)x^3+(A+B)x^2+Bx}{x^3(x+1)} = \dfrac{(A+C)x^2+(A+B)x+B}{x^2(x+1)}.$
    Thus, $B = 2$, $A = 1$, and $C = 3$.

    (d) $\displaystyle\int p(x)\,dx = \int \left(\dfrac{1}{x} + \dfrac{2}{x^2} + \dfrac{3}{1+x}\right) dx = \ln|x| - 2/x + 3\ln|1+x| + C.$

7. (a) f is the quotient of two polynomials.

    (b) Both the numerator and the denominator of the expression defining f have degree 2.

    (c) $f(x) = \dfrac{x^2+1}{(x-1)^2} = \dfrac{x^2+1-2x+2x}{x^2-2x+1} = 1 + \dfrac{2x}{x^2-2x+1}.$

(d) $\displaystyle\int f(x)\,dx = \int\left(1 + \frac{2x}{x^2 - 2x + 1}\right)dx = \int\left(1 + \frac{2x - 2}{x^2 - 2x + 1} + \frac{2}{(x-1)^2}\right)dx$

$\displaystyle = x + \ln(x^2 - 2x + 1) - \frac{2}{x - 1} = x + 2\ln|x - 1| - \frac{2}{x - 1}.$

8. (a) The expression defining g is the ratio of two polynomials.

(b) No, because the degree of the polynomial in the numerator is greater than the degree of the polynomial in the denominator.

(c) $\displaystyle g(x) = \frac{(x+1)^3}{x^2 + x + 1} = \frac{x^3 + 3x^2 + 3x + 1}{x^2 + x + 1} = \frac{x(x^2 + x + 1) + 2(x^2 + x + 1) - 1}{x^2 + x + 1}$

$\displaystyle = x + 2 - \frac{1}{x^2 + x + 1}.$

(d) $\displaystyle\int g(x)\,dx = \int\left(x + 2 - \frac{1}{x^2 + x + 1}\right)dx = \int\left(x + 2 - \frac{1}{(x + 1/2)^2 + 3/4}\right)dx$

$\displaystyle = \frac{x^2}{2} + 2x - \frac{2}{\sqrt{3}}\arctan\big((1 + 2x)/\sqrt{3}\big).$

9. Yes, because $x^2 + 2x + 3$ is an irreducible quadratic polynomial.

10. $\displaystyle g(x) = \frac{A}{x + 1} + \frac{Bx + C}{x^2 - 3x + 4}.$

11. No — $x^2 - 2x + 1 = (x - 1)^2$ so the partial fraction decomposition of g has the form
$\displaystyle\frac{A}{x + 1} + \frac{B}{x - 1} + \frac{C}{(x - 1)^2}.$

12. No — $x^2 + 2x + 1 = (x + 1)^2$ so $q(x) = (x + 1)^3$. Thus, the partial fraction decomposition
of g has the form $\displaystyle\frac{A}{x + 1} + \frac{B}{(x + 1)^2} + \frac{C}{(x + 1)^3}.$

13. No — since $(x + 1)^2$ is a repeated linear factor and $x^2 + 1$ is an irreducible quadratic factor,
the partial fraction decomposition of h has the form $\displaystyle\frac{A}{x + 1} + \frac{B}{(x + 1)^2} + \frac{Cx + D}{x^2 + 1}.$

14. No — the partial fraction decomposition of k is $\displaystyle\frac{A}{x + 1} + \frac{Bx + C}{x^2 + 1} + \frac{Dx + E}{(x^2 + 1)^2} + \frac{Fx + G}{(x^2 + 1)^3}.$

15. (a) After multiplying, the result is $\displaystyle\frac{x^2 + 3x - 1}{(x + 1)(x - 2)} = A + \frac{Bx}{x + 1} + \frac{Cx}{x - 2}.$ Substituting
$x = 0$ into this equation yields $A = 1/2.$

(b) After multiplying, the result is $\displaystyle\frac{x^2 + 3x - 1}{x(x - 2)} = \frac{A(x + 1)}{x} + B + \frac{C(x + 1)}{x - 2}.$
Substituting $x = -1$ into this equation yields $B = -1.$

(c) After multiplying, the result is $\displaystyle\frac{x^2 + 3x - 1}{x(x + 1)} = \frac{A(x - 2)}{x} + \frac{B(x - 2)}{x + 1} + C.$
Substituting $x = 2$ into this equation yields $C = 3/2.$

(d) $\int \dfrac{x^2 + 3x - 1}{x(x+1)(x-2)}\, dx = \dfrac{1}{2} \int \left(\dfrac{1}{x} - \dfrac{2}{x+1} + \dfrac{3}{x-2} \right) dx$

$\qquad\qquad\qquad\qquad = \left(\ln |x| - 2 \ln |1 + x| + 3 \ln |x - 2| \right)/2 + C.$

16. (a) $\dfrac{6}{(x-2)(x^2-1)} = \dfrac{6}{(x-2)(x-1)(x+1)} = \dfrac{2}{x-2} - \dfrac{3}{x-1} + \dfrac{1}{x+1}.$

    (b) $\int \dfrac{6}{(x-2)\left(x^2-1\right)}\, dx = 2 \ln |x - 2| - 3 \ln |x - 1| + \ln |x + 1| + C.$

17. Since $\dfrac{x^2 + 2x + 5}{(x-1)(x+1)(x+2)} = \dfrac{4}{3(x-1)} - \dfrac{2}{x+1} + \dfrac{5}{3(x+2)},$

    $\int \dfrac{x^2 + 2x + 5}{(x-1)(x+1)(x+2)}\, dx = \dfrac{4 \ln |x-1|}{3} - 2 \ln |x+1| + \dfrac{5 \ln |x+2|}{3} + C.$

18. $\int \dfrac{x^2 + 2x + 5}{x(x^2-4)}\, dx = \int \left(-\dfrac{5}{4x} + \dfrac{13}{8(x-2)} + \dfrac{5}{8(x+2)} \right) dx$

    $\qquad\qquad\qquad\quad = -\dfrac{5 \ln |x|}{4} + \dfrac{13 \ln |x-2|}{8} + \dfrac{5 \ln |x+2|}{8} + C.$

19. (a) After multiplying, the result is $\dfrac{x^2-1}{x^2+4} = A + \dfrac{x(Bx+C)}{x^2+4}$. Substituting $x = 0$ into this equation yields $A = -1/4$.

    (b) Using the result found in part (a), the partial fraction decomposition equation becomes $\dfrac{x^2-1}{x(x^2+4)} = -\dfrac{1}{4x} + \dfrac{Bx+C}{x^2+4}$. Substituting $x = 1$ into this equation yields $0 = -\dfrac{1}{4} + \dfrac{B+C}{5}$. Thus, $B + C = 5/4.$

    (c) Using the result found in part (a), the partial fraction decomposition equation becomes $\dfrac{x^2-1}{x(x^2+4)} = -\dfrac{1}{4x} + \dfrac{Bx+C}{x^2+4}$. Substituting $x = -1$ into this equation yields $0 = \dfrac{1}{4} + \dfrac{C-B}{5}$. Thus, $B - C = 5/4.$

    (d) $B = 5/4$ and $C = 0$ is the unique solution of the system equations $B + C = 5/4$ and $B - C = 5/4.$

    (e) $\int \dfrac{x^2-1}{x\left(x^2+4\right)}\, dx = \dfrac{1}{4} \int \left(-\dfrac{1}{x} + \dfrac{5x}{x^2+4} \right) dx = -\dfrac{\ln |x|}{4} + \dfrac{5 \ln \left| x^2+4 \right|}{8} + C.$

20. $\dfrac{3x^2 + 7x + 5}{(x+1)(x^2+2x+2)} = \dfrac{1}{1+x} + \dfrac{2x+3}{x^2+2x+2}.$ Thus,

    $\int \dfrac{3x^2 + 7x + 5}{(x+1)(x^2+2x+2)}\, dx = \ln |1 + x| + \ln(x^2 + 2x + 2) + \arctan(x+1) + C.$

21. Since $\left(\dfrac{4}{1-x} + 5 \ln |3 + x| + C \right)' = \dfrac{4}{(1-x)^2} + \dfrac{5}{3+x} = \dfrac{17 - 6x + 5x^2}{(1-x)^2(3+x)},$

    $q(x) = 17 - 6x + 5x^2.$

22. No. Since $q(x)$ is a quadratic polynomial, the partial fraction decomposition of the function

$f(x) = \dfrac{q(x)}{(1 - x^2)(3 + x)}$ has the form $\dfrac{A}{1 - x} + \dfrac{B}{1 + x} + \dfrac{C}{3 + x}$. Thus,

$\displaystyle\int f(x)\,dx = -A \ln|1 - x| + B \ln|1 + x| + C \ln|3 + x| + \text{constant}$. This implies that

there is no quadratic polynomial with the desired property.

23. No, the antiderivative of a rational function never involves the arcsine function.

24. No — the partial fraction decomposition of $\dfrac{q(x)}{(1 - x)^2(x + 3)} = \dfrac{A}{1 - x} + \dfrac{B}{(1 - x)^2} + \dfrac{C}{x + 3}$,

so $\displaystyle\int \dfrac{q(x)}{(1 - x)^2(x + 3)}\,dx = A \ln|1 - x| + \dfrac{B}{1 - x} + C \ln|x + 3| + D$.

25. (a) After multiplying, the result is $x^2 = A(x + 1)^2 + B(x + 1) + C$. Substituting $x = -1$ into this equation yields $C = 1$.

(b) Since $C = 1$, the partial fraction decomposition equation is

$\dfrac{x^2}{(x + 1)^3} = \dfrac{A}{x + 1} + \dfrac{B}{(x + 1)^2} + \dfrac{1}{(x + 1)^3}$. Substituting $x = 0$ into this equation

yields $0 = A + B + 1$. Thus, $A + B = -1$.

(c) Substituting $C = 1$ and $x = 1$ into the partial fraction decomposition equation yields

$\dfrac{1}{8} = \dfrac{A}{2} + \dfrac{B}{4} + \dfrac{1}{8}$. Thus, $2A + B = 0$.

(d) $A = 1$ and $B = -2$ is the unique solution of the system of equations $A + B = -1$ and $2A + B = 0$.

(e) $\displaystyle\int \dfrac{x^2}{(x + 1)^3}\,dx = \int \left(\dfrac{1}{x + 1} - \dfrac{2}{(x + 1)^2} + \dfrac{1}{(x + 1)^3} \right)\,dx$

$\qquad\qquad\qquad = \ln|x + 1| + \dfrac{2}{x + 1} - \dfrac{1}{2(x + 1)^2} + C$.

26. $\dfrac{2 - x}{x^2(x + 2)} = -\dfrac{1}{x} + \dfrac{1}{x^2} + \dfrac{1}{x + 2}$ so $\displaystyle\int \dfrac{2 - x}{x^2(x + 2)}\,dx = -\ln|x| - \dfrac{1}{x} + \ln|x + 2| + C$.

27. (a) The partial fraction decomposition of $1/p(x)$ has the form $\dfrac{1}{p(x)} = \dfrac{A}{x + 1} + \dfrac{B}{x - 2}$.

Solving this equation yields $A = -1/3$ and $B = 1/3$.

(b) The partial fraction decomposition of $x/p(x)$ has the form $\dfrac{x}{p(x)} = \dfrac{A}{x + 1} + \dfrac{B}{x - 2}$.

Solving this equation yields $A = 1/3$ and $B = 2/3$.

(c) Parts (a) and (b) imply that $\dfrac{4 - 3x}{(x + 1)(x - 2)} = -\dfrac{2}{3(x - 2)} - \dfrac{7}{3(x + 1)}$. Thus,

$$\int \dfrac{4 - 3x}{(x + 1)(x - 2)}\,dx = -\dfrac{1}{3} \int \left(\dfrac{2}{x - 2} + \dfrac{7}{x + 1} \right)\,dx$$

$$= -\dfrac{2 \ln|x - 2|}{3} - \dfrac{7 \ln|x + 1|}{3} + C.$$

28. (a) The partial fraction decomposition of $1/p(x)$ has the form $\dfrac{1}{p(x)} = \dfrac{A}{x-2} + \dfrac{B}{x+3}$.
   Solving this equation yields $A = 1/5$ and $B = -1/5$.

   (b) The partial fraction decomposition of $x/p(x)$ has the form $\dfrac{x}{p(x)} = \dfrac{A}{x-2} + \dfrac{B}{x+3}$.
   Solving this equation yields $A = 2/5$ and $B = 3/5$.

   (c) Parts (a) and (b) imply that $\dfrac{4x+5}{(x-2)(x+3)} = \dfrac{13}{5(x-2)} + \dfrac{7}{5(x+3)}$. Thus,
   $$\int \dfrac{4x+5}{(x-2)(x+3)}\, dx = \dfrac{13\ln|x-2|}{5} + \dfrac{7\ln|x+3|}{5} + C.$$

29. (a) The partial fraction decomposition of $1/p(x)$ has the form $\dfrac{1}{p(x)} = \dfrac{A}{x-1} + \dfrac{Bx+C}{x^2+1}$.
   Solving this equation yields $A = 1/2$, $B = -1/2$, and $C = -1/2$.

   (b) The partial fraction decomposition of $x/p(x)$ has the form $\dfrac{x}{p(x)} = \dfrac{A}{x-1} + \dfrac{Bx+C}{x^2+1}$.
   Solving this equation yields $A = 1/2$, $B = -1/2$, and $C = 1/2$.

   (c) The partial fraction decomposition of $x^2/p(x)$ has the form $\dfrac{x^2}{p(x)} = \dfrac{A}{x-1} + \dfrac{Bx+C}{x^2+1}$.
   Solving this equation yields $A = 1/2$, $B = 1/2$, and $C = 1/2$.

   (d) Parts (a)–(c) imply that $\dfrac{2+3x+4x^2}{(x-1)(x^2+1)} = \dfrac{9}{2(x-1)} + \dfrac{5-x}{2(x^2+1)}$. Thus,
   $$\int \dfrac{2+3x+4x^2}{(x-1)(x^2+1)}\, dx = \dfrac{9\ln|x-1|}{2} + \dfrac{5\arctan x}{2} - \dfrac{\ln(x^2+1)}{4} + C.$$

30. (a) $\dfrac{1}{p(x)} = \dfrac{1}{5(x+2)} - \dfrac{x}{5(x^2+2x+5)}$.

   (b) $\dfrac{x}{p(x)} = -\dfrac{2}{5(x+2)} + \dfrac{2x+5}{5(x^2+2x+5)}$.

   (c) $\dfrac{x^2}{p(x)} = \dfrac{4}{5(x+2)} + \dfrac{x-10}{5(x^2+2x+5)}$.

   (d) $\displaystyle\int \dfrac{3-4x+5x^2}{(x+2)(x^2+2x+5)}\, dx = \dfrac{31\ln|x+2|}{5}$
   $$- \dfrac{3\ln(x^2+2x+5)}{5} - \dfrac{32\arctan\big((x+1)/2\big)}{5}.$$

31. The partial fraction decomposition of the integrand has the form
$$\dfrac{1}{(x-2)(x^2+1)} = \dfrac{A}{x-2} + \dfrac{Bx+C}{x^2+1}.$$

To determine the constants, multiply both sides of the equation above by $(x-2)(x^2+1)$ to obtain
$$1 = A(x^2+1) + (Bx+C)(x-2) = (A+B)x^2 + (C-2B)x + (A-2C).$$

It follows that $A + B = 0$, $C - 2B = 0$, and $A - 2C = 1$. Inserting $x = 2$ into the equation displayed above, we find that $A = 1/5$ and, therefore, $B = -1/5$ and $C = -2/5$. Thus,

$$\int_0^1 \frac{dx}{(x-2)(x^2+1)} = \frac{1}{5}\int_0^1 \frac{dx}{x-2} - \frac{1}{5}\int_0^1 \frac{x}{x^2+1}\,dx - \frac{2}{5}\int_0^1 \frac{dx}{x^2+1}$$

$$= \left(\frac{1}{5}\ln|x-2| - \frac{1}{10}\ln\left|x^2+1\right| - \frac{2}{5}\arctan x\right)\Big]_0^1$$

$$= \left(-\frac{1}{10}\ln 2 - \frac{2}{5}\frac{\pi}{4}\right) - \frac{1}{5}\ln 2 = -\frac{3}{10}\ln 2 - \frac{\pi}{10}.$$

32. $\displaystyle \int \frac{dx}{x^3+1} = \int \frac{dx}{(x+1)(x^2-x+1)} = \frac{1}{3}\int\left(\frac{1}{x+1} + \frac{2-x}{x^2-x+1}\right)dx$

$$= \frac{1}{3}\ln|x+1| - \frac{1}{6}\ln\left|x^2-x+1\right| + \frac{1}{2}\int \frac{dx}{x^2-x+1}$$

$$= \frac{1}{3}\ln|x+1| - \frac{1}{6}\ln\left|x^2-x+1\right| + \frac{1}{\sqrt{3}}\arctan\left(\frac{2x-1}{\sqrt{3}}\right).$$

Thus, $\displaystyle \int_0^2 \frac{dx}{x^3+1} = \frac{1}{6}\left(\ln 3 + \sqrt{3}\pi\right).$

33. $\displaystyle \int \frac{2x+1}{(x-2)(x+3)}\,dx = \int\left(\frac{1}{x-2} + \frac{1}{x+3}\right)dx = \ln|x-2| + \ln|x+3| + C.$

34. $\displaystyle \int \frac{x+1}{(x-1)(x+2)}\,dx = \int\left(\frac{2}{3}\cdot\frac{1}{x-1} + \frac{1}{3}\cdot\frac{1}{x+2}\right)dx$

$$= \frac{2}{3}\ln|x-1| + \frac{1}{3}\ln|(x+2| + C.$$

35. $\displaystyle \int \frac{5x^2+3x-2}{x^3+2x^2}\,dx = \int\left(-\frac{1}{x^2} + \frac{2}{x} + \frac{3}{x+2}\right)dx = \frac{1}{x} + 2\ln|x| + 3\ln|x+2| + C.$

36. Let $I = \displaystyle\int \frac{4x^2-3x+2}{x(2x-1)^2}\,dx$. Use partial fractions:

$\displaystyle \frac{4x^2-3x+2}{x(2x-1)^2} = \frac{A}{x} + \frac{B}{(2x-1)} + \frac{C}{(2x-1)^2}$. Solving gives $A = 2$, $B = -2$, and $C = 3$.

Therefore, $I = 2\ln|x| - \ln|2x-1| - \dfrac{3}{2}\cdot\dfrac{1}{(2x-1)} + C.$

37. $\displaystyle \int \frac{x^4}{x^4-1}\,dx = \int\left(1 + \frac{1}{4}\cdot\frac{1}{x-1} - \frac{1}{4}\cdot\frac{1}{x+1} - \frac{1}{2}\cdot\frac{1}{1+x^2}\right)dx$

$$= x + \frac{1}{4}\ln|x-1| - \frac{1}{4}\ln|x+1| - \frac{1}{2}\arctan x + C.$$

38. $\displaystyle \int \frac{x^3}{x^2+1}\,dx = \int\left(x - \frac{x}{x^2+1}\right)dx = \frac{1}{2}x^2 - \frac{1}{2}\ln(x^2+1) + C.$

39. $\displaystyle\int \frac{x^3}{x^2-1}\,dx = \int \left(x + \frac{1}{2}\cdot\frac{1}{x-1} + \frac{1}{2}\cdot\frac{1}{x+1}\right)dx$

$\displaystyle\qquad = \frac{1}{2}x^2 + \frac{1}{2}\ln|x-1| + \frac{1}{2}\ln|x+1| + C.$

40. $\displaystyle\int \frac{3x^2-1}{(x-1)(x+2)}\,dx = \int \left(3 + \frac{2}{3}\cdot\frac{1}{x-1} - \frac{11}{3}\cdot\frac{1}{x+2}\right)dx$

$\displaystyle\qquad = 3x + \frac{2}{3}\ln|x-1| - \frac{11}{3}\ln|x+2| + C.$

41. $\displaystyle\int \frac{dx}{x\sqrt{x+1}} = 2\int \frac{du}{u^2-1} = \int\left(\frac{1}{u-1} - \frac{1}{u+1}\right)du$

$\displaystyle\qquad = \ln\left|\frac{u-1}{u+1}\right| + C = \ln\left|\frac{\sqrt{x+1}-1}{\sqrt{x+1}+1}\right| + C.$

42. $\displaystyle\int \frac{dx}{x\sqrt{x-1}} = 2\int \frac{du}{u^2+1} = 2\arctan u + C = 2\arctan(\sqrt{x-1}) + C.$

43. (a) Since $\left(\ln|x+c|\right)' = \dfrac{1}{x+c}$, $\displaystyle\int \frac{dx}{x+c} = \ln|x+c|.$

(b) Since $\left(\dfrac{1}{1-n}\dfrac{1}{(x+c)^{n-1}}\right)' = \dfrac{1}{(x+c)^n}$, $\displaystyle\int \frac{dx}{(x+c)^n} = \frac{1}{(1-n)(x+c)^{n-1}}.$

(c) Using the substitution $u = ax$ in parts (a) and (b) (and replacing the constant c by b) leads to the desired result.

44. $\left(\dfrac{1}{|d|}\arctan\left(\dfrac{x}{|d|}\right)\right)' = \dfrac{1}{|d|^2}\dfrac{1}{(x/|d|)^2 + 1} = \dfrac{1}{x^2+d^2}.$

45. Let $u = x^2 + d^2$. Then, $\displaystyle\int \frac{x}{x^2+d^2}\,dx = \frac{1}{2}\int \frac{du}{u} = \frac{1}{2}\ln|u| = \frac{1}{2}\ln\left(x^2+d^2\right).$

46. (a) Integration by parts with $u = 1/(x^2+d^2)^n$ and $dv = dx$ leads directly to the desired identity.

(b) Using the algebraic identity $x^2 = (x^2+d^2) - d^2$ in the result from part (a) leads to

$$\int \frac{x^2}{\left(x^2+d^2\right)^{n+1}}\,dx = \int \frac{(x^2+d^2)-d^2}{\left(x^2+d^2\right)^{n+1}}\,dx = \int \frac{dx}{\left(x^2+d^2\right)^n} - d^2\int \frac{dx}{\left(x^2+d^2\right)^{n+1}}.$$

Rearranging the terms of this equation leads to the new equation

$$2nd^2\int \frac{dx}{\left(x^2+d^2\right)^{n+1}} = \frac{x}{\left(x^2+d^2\right)^n} + (2n-1)\int \frac{dx}{\left(x^2+d^2\right)^n}.$$

The desired identity follows from dividing both sides of this equation by $2nd^2$.

47. Let $u = x^2 + d^2$. Then,

$$\int \frac{x}{\left(x^2 + d^2\right)^n}\, dx = \frac{1}{2} \int \frac{du}{u^n} = \frac{1}{2(1-n)u^{n-1}} = \frac{1}{2(1-n)\left(x^2 + d^2\right)^{n-1}}.$$

§8.3 Trigonometric Antiderivatives

1. (a) $\int \cos x \, dx = \sin x + C.$

    (b) $\int \sin x \, dx = -\cos x + C.$

    (c) $\int \sec^2 x \, dx = \tan x + C.$

    (d) $\int \sec x \tan x \, dx = \sec x + C.$

2. (a) $\int \sin x \cos x \, dx = \frac{1}{2} \sin^2 x + C.$

    (b) $\int \tan x \, dx = \ln |\sec x| + C.$

    (c) $\int \sec^2 x \tan x \, dx = \frac{1}{2} \tan^2 x + C.$

3. No—the two answers are equal. To see this, use the identity $\cos^2 x = 1 - \sin^2 x.$

4. (a) $\int \sin^4 x \, dx = -\dfrac{\sin^3 x \cos x}{4} + \dfrac{3}{4} \int \sin^2 x \, dx$

    $$= -\frac{\sin^3 x \cos x}{4} - \frac{3}{8} \cos x \sin x + \frac{3x}{8} + C.$$

    (b) Yes, the two answers can be shown to be the same using trigonometric identities.

5. (a) $\int \cos^3 x \sin^4 x \, dx = \int (1 - u^2) u^4 \, du = u^5/5 - u^7/7 = \dfrac{\sin^5 x}{5} - \dfrac{\sin^7 x}{7} + C.$

    (b) $\int \cos^3 x \sin^4 x \, dx = \int \cos^3 x \left(1 - \cos^2 x\right)^2 dx = \int \left(\cos^3 x - 2\cos^5 x + \cos^7 x\right) dx$

    $$= \int \cos^3 x \, dx - 2 \int \cos^5 x \, dx + \frac{\cos^6 x \sin x}{7} + \frac{6}{7} \int \cos^5 x \, dx$$

    $$= \int \cos^3 x \, dx - \frac{8\cos^4 x \sin x}{35} - \frac{32}{35} \int \cos^3 x \, dx + \frac{\cos^6 x \sin x}{7}$$

    $$= \frac{\cos^2 x \sin x}{35} + \frac{2}{35} \int \cos x \, dx - \frac{8\cos^4 x \sin x}{35} + \frac{\cos^6 x \sin x}{7}$$

    $$= \frac{\cos^2 x \sin x}{35} + \frac{2 \sin x}{35} - \frac{8\cos^4 x \sin x}{35} + \frac{\cos^6 x \sin x}{7} + C.$$

    (c) Yes, trigonometric identities can be used to show that the two expressions define the same function.

6. $\int \sec^3 x \tan^3 x \, dx = \int u^2(u^2 - 1) \, du = u^5/5 - u^3/3 = \dfrac{\sec^5 x}{5} - \dfrac{\sec^3}{3} + C.$

7. (a) $\cos t = \sqrt{a^2 - x^2}/a$ [adjacent/hypotenuse].

   (b) $\tan t = x/\sqrt{a^2 - x^2}$ [opposite/adjacent].

   (c) $\sin(2t) = 2\sin t \cos t = 2x\sqrt{a^2 - x^2}/a^2$.

8. (a) $\cos t = a/\sqrt{a^2 + x^2}$ [adjacent/hypotenuse].

   (b) $\sin t = x/\sqrt{a^2 + x^2}$ [opposite/hypotenuse].

   (c) $\sec(2t) = 1/\cos(2t) = \dfrac{1}{1 - 2\sin^2 t} = \dfrac{1}{1 - 2x^2/(a^2 + x^2)} = \dfrac{a^2 + x^2}{a^2 - x^2}.$

9. $\displaystyle\int \frac{dx}{\sqrt{x^2 + 2x + 5}} = \int \frac{du}{\sqrt{u^2 + 4}} = \ln\left|\sqrt{u^2 + 4} + u\right|$

   $\qquad\qquad = \ln\left|\sqrt{x^2 + 2x + 5} + x + 1\right| + C.$

10. $\displaystyle\int_{-1}^{1} \frac{dx}{\sqrt{x^2 + 2x + 5}} = \int_{-1}^{1} \frac{dx}{\sqrt{(x+1)^2 + 4}} = \int_{0}^{2} \frac{du}{\sqrt{u^2 + 4}} = \ln(\sqrt{2} + 1).$ (See Example 8.)

11. $\displaystyle\int \sin^2(3x)\, dx = \frac{1}{2}\int (1 - \cos(6x))\, dx = \frac{1}{2}x - \frac{1}{12}\sin(6x) + C.$

12. $\displaystyle\int \cos^2(x/3)\, dx = \frac{1}{2}\int \left(1 + \cos(2x/3)\right) dx = \frac{1}{2}x + \frac{3}{4}\sin(2x/3) + C.$

13. Let $u = \sin x$. Then, $\displaystyle\int \sin^2 x \cos x\, dx = \int u^2\, du = u^3/3 = \sin^3 x/3 + C.$

14. Let $u = \sin x$. Then, $du = \cos x\, dx$ and

   $\displaystyle\int \sin^3 x \cos^3 x\, dx = \int \sin^3 x \cos^2 x \cos x\, dx = \int \sin^3 x (1 - \sin^2 x) \cos x\, dx$

   $\qquad\qquad = \int \sin^3 x \cos x\, dx - \int \sin^5 x \cos x\, dx = \int u^3\, du - \int u^5\, du$

   $\qquad\qquad = \frac{1}{4}u^4 - \frac{1}{6}u^6 + C = \frac{1}{4}\sin^4 x - \frac{1}{6}\sin^6 x + C.$

15. $\displaystyle\int \cos^2 x \sin^3 x\, dx = \int (\cos^2 x - \cos^4 x) \sin x\, dx = -\frac{1}{3}\cos^3 x + \frac{1}{5}\cos^5 x + C.$

16. $\displaystyle\int \cos^3(2x) \sin^2(2x)\, dx = \int \cos(2x)\left(\sin^2(2x) - \sin^4(2x)\right) dx$

   $\qquad\qquad = \frac{1}{6}\sin^3(2x) - \frac{1}{10}\sin^5(2x) + C.$

17. $\displaystyle\int \sin^2 x \cos^2 x\, dx = \int (\sin^2 x - \sin^4 x)\, dx = \frac{1}{8}x + \frac{1}{8}\cos x \sin x - \frac{1}{4}\cos^3 x \sin x + C.$

18. $\displaystyle\int \cos^4 x \sin^2 x \, dx = \int \left(\cos^4 x - \cos^6 x\right) dx$

$$= \frac{1}{16}x + \frac{1}{16}\cos x \sin x + \frac{1}{24}\cos^3 x \sin x - \frac{1}{6}\sin x \cos^5 x + C.$$

19. $\displaystyle\int \tan^4 x \, dx = \frac{1}{3}\tan^3 x - \tan x + x + C.$

20. Using the reduction formula,

$$\int \sec^4(3x)\, dx = \frac{\sec^2(3x)\tan(3x)}{9} + \frac{2}{9}\int \sec^2(3x)\, dx = \frac{\sec^2(3x)\tan(3x)}{9} + \frac{2}{27}\tan(3x) + C.$$

21. $\displaystyle\int \sec^2 x \tan^2 x \, dx = \frac{1}{3}\tan^3 x + C.$

22. Let $u = \tan x$. Then, $\displaystyle\int \sec^2 x \tan^4 x \, dx = \int u^4 \, du = u^5/5 = \tan^5 x/5 + C.$

23. $\displaystyle\int \sec x \tan^2 x \, dx = \int \sec x (\sec^2 x - 1) \, dx = \int \sec^3 x \, dx - \int \sec x \, dx$

$$= \frac{1}{2}\left(\sec x \tan x + \int \sec x \, dx\right) - \int \sec x \, dx$$

$$= \frac{1}{2}\left(\tan x \sec x - \ln|\sec x + \tan x|\right) + C.$$

24. $\displaystyle\int \sec^3 x \tan^2 x \, dx = \int \left(\sec^5 x - \sec^3 x\right) dx$

$$= \frac{1}{4}\sec^3 x \tan x - \frac{1}{8}\sec x \tan x - \frac{1}{8}\ln(\sec x + \tan x) + C.$$

25. $\displaystyle\int \sec x \tan^3 x \, dx = \int \sec x \tan x \tan^2 x \, dx = \int \sec^2 x \sec x \tan x \, dx - \int \sec x \tan x \, dx$

$$= \sec^3 x/3 - \sec x + C.$$

26. $\displaystyle\int \frac{dx}{x^2\sqrt{x^2+1}} = \int \frac{\cos t}{\sin^2 t}\, dt = -\frac{1}{\sin t} + C = -\frac{\sqrt{x^2+1}}{x} + C$

$[x = \tan t, dx = \sec^2 t \, dt, \sin t = x/\sqrt{x^2+1}].$

27. $\displaystyle\int \frac{dx}{x^2\sqrt{4-x^2}} = \frac{1}{4}\int \csc^2 t \, dt = -\frac{1}{4}\cot t + C = -\frac{1}{4}\frac{\cos t}{\sin t} + C = -\frac{\sqrt{4-x^2}}{4x} + C$

$[x = 2\sin t, dx = 2\cos t \, dt, \cos t = \sqrt{1-x^2/4}].$

28. $\displaystyle\int \frac{dx}{\sqrt{1+x^2}} = \int \frac{\sec^2 t}{\sqrt{1+\tan^2 t}}\, dt = \int \sec t \, dt$

$$= \ln|\tan t + \sec t| + C = \ln\left|x + \sqrt{1+x^2}\right| + C.$$

29. $\int \sqrt{1-x^2}\,dx = \int \sqrt{1-\sin^2 t}\,\cos t\,dt = \int \cos^2 t\,dt = \dfrac{1}{2}\cos t\sin t + \dfrac{1}{2}t + C$

$\quad\quad = \dfrac{1}{2}x\sqrt{1-x^2} + \dfrac{1}{2}\arcsin x + C.$

30. $\int \dfrac{dx}{(x^2+4)^2} = \dfrac{1}{8}\int \dfrac{\sec^2 t}{(1+\tan^2 t)^2}\,dt = \dfrac{1}{8}\int \cos^2 t\,dt = \dfrac{t}{16} + \dfrac{\sin(2t)}{32} + C$

$\quad\quad = \dfrac{t}{16} + \dfrac{\sin t\cos t}{16} + C = \dfrac{1}{16}\arctan(x/2) + \dfrac{x}{8(4+x^2)} + C$

$[x = 2\tan t,\, dx = 2\sec^2 t\,dt,\, \sin t = x/\sqrt{4+x^2},\, \cos t = 2/\sqrt{4+x^2}].$

31. $\int x^2\sqrt{1-x^2}\,dx = \int \sin^2 t\cos^2 t\,dt = \int (\cos^2 t - \cos^4 t)\,dt$

$\quad\quad = \dfrac{t}{8} + \dfrac{1}{16}\sin(2t) - \dfrac{1}{4}\cos^3 t\sin t + C$

$\quad\quad = \dfrac{1}{8}\arcsin x + \dfrac{1}{8}x\sqrt{1-x^2} - \dfrac{1}{4}x\left(1-x^2\right)^{3/2} + C$

$[x = \sin t,\, dx = \cos t\,dt,\, \cos t = \sqrt{1-x^2}].$

32. $\int \dfrac{x^2}{\sqrt{9-x^2}}\,dx = 9\int \sin^2 t\,dt = \dfrac{9t}{2} - \dfrac{9}{4}\sin(2t) + C = \dfrac{9t}{2} - \dfrac{9}{2}\sin t\cos t + C$

$\quad\quad = \dfrac{9}{2}\arcsin(x/3) - \dfrac{1}{2}x\sqrt{9-x^2} + C$

$[x = 3\sin t,\, dx = 3\cos t\,dt,\, \cos t = \sqrt{1-x^2/9}].$

33. (a) $\cos t = 1/\sec t = a/x$ [adjacent/hypotenuse].

$\quad$ (b) $\sin t = \sqrt{x^2-a^2}/x$ [opposite/hypotenuse].

$\quad$ (c) $\tan(2t) = \dfrac{2\tan t}{1-\tan^2 t} = \dfrac{2\tan t}{2-\sec^2 t} = \dfrac{2a\sqrt{x^2-a^2}}{2a^2-x^2}.$

34. The absolute value is required because $\sqrt{x^2-4} > 0$ when $x < -2$, but $2\tan t < 0$ when $\pi/2 < t < \pi$.

35. $\int \dfrac{dx}{x^2\sqrt{x^2-4}} = \dfrac{1}{4}\int \cos t\,dt = \dfrac{1}{4}\sin t + C = \dfrac{\sqrt{x^2-4}}{4x} + C$

$[x = 2\sec t,\, dx = 2\sec t\tan t\,dt,\, \sin t = \sqrt{1-4/x^2}].$

36. Note that $\displaystyle\int_{-4}^{-3} \dfrac{dx}{x^2\sqrt{x^2-4}} = \int_{3}^{4} \dfrac{dx}{x^2\sqrt{x^2-4}}$. Also, using the substitution $x = 2\sec t$,

$\int \dfrac{dx}{x^2\sqrt{x^2-4}} = \dfrac{1}{4}\int \cos t\,dt = (\sin t)/4 = \dfrac{\sqrt{x^2-4}}{4x}$ so

$\displaystyle\int_{-4}^{-3} \dfrac{dx}{x^2\sqrt{x^2-4}} = \dfrac{\sqrt{12}}{16} - \dfrac{\sqrt{5}}{12} = \dfrac{\sqrt{3}}{8} - \dfrac{\sqrt{5}}{12}.$

37. Let $x = \sec t$. When $x > 0$, $\sqrt{x^2 - 1} = \tan t$ (since $\tan t > 0$). Thus,

$$\int_1^2 \frac{\sqrt{x^2 - 1}}{x}\, dx = \int_0^{\pi/3} \tan^2 t\, dt = \tan t - t \Big]_0^{\pi/3} = \sqrt{3} - \pi/3.$$

38. Let $x = \sec t$. When $x < 0$, $\sqrt{x^2 - 1} = -\tan t$ (since $\tan t < 0$). Thus,

$$\int_{-2}^{-1} \frac{\sqrt{x^2 - 1}}{x}\, dx = -\int_{2\pi/3}^{\pi} \tan^2 t\, dt = t - \tan t \Big]_{2\pi/3}^{\pi} = \pi/3 - \sqrt{3}.$$

39. $\displaystyle \int \frac{\sin^3 x}{\cos x}\, dx = \int \frac{(1 - \cos^2 x) \sin x}{\cos x}\, dx = \frac{1}{2} \cos^2 x - \ln|\cos x| + C.$

40. $\displaystyle \int \sin(2x) \cos^2 x\, dx = \frac{1}{2} \int \sin(2x)\big(1 + \cos(2x)\big)\, dx = -\frac{1}{4} \cos(2x) + \frac{1}{4} \cos^2(2x) + C.$

41. $\displaystyle \int \sqrt{\cos x}\, \sin^5 x\, dx = \int \sqrt{\cos x}\, \big(1 - \cos^2 x\big)^2 \sin x\, dx$

$$= -\frac{2}{3} (\cos x)^{3/2} + \frac{4}{7} (\cos x)^{7/2} - \frac{2}{11} (\cos x)^{11/2} + C.$$

42. $\displaystyle \int \sqrt{1 + \sin x}\, dx = \int \sqrt{1 + \sin x} \cdot \frac{\sqrt{1 - \sin x}}{\sqrt{1 - \sin x}}\, dx$

$$= \int \frac{\cos x}{\sqrt{1 - \sin x}}\, dx = -2\sqrt{1 - \sin x} + C.$$

43. $\displaystyle \int \sqrt{1 + x^2}\, dx = \int \sec t \sec^2 t\, dt = \int \sec^3 t\, dt = \frac{1}{2} \tan t \sec t + \frac{1}{2} \ln|\sec t + \tan t| + C$

$$= \frac{1}{2} x \sqrt{1 + x^2} + \frac{1}{2} \ln\left|\sqrt{1 + x^2} + x\right| + C$$

$[x = \tan t, dx = \sec^2 t\, dt, \sec t = \sqrt{1 + x^2}].$

44. $\displaystyle \int \frac{\sqrt{4 - x^2}}{x^2}\, dx = \int \frac{\cos^2 t}{\sin^2 t}\, dt = \int \frac{1 - \sin^2 t}{\sin^2 t}\, dt = \int \left(\csc^2 t - 1\right)\, dt$

$$= -\cot t - t + C = -\frac{\cos t}{\sin t} - t + C = -\frac{\sqrt{4 - x^2}}{x} - \arcsin(x/2) + C.$$

$[x = 2 \sin t, dx = 2 \cos t\, dt, \cos t = \sqrt{1 - x^2/4}].$

45. First, note that

$$\int \frac{x + 2}{x\left(x^2 + 1\right)}\, dx = \int \frac{dx}{x^2 + 1} + 2 \int \frac{dx}{x\left(x^2 + 1\right)} = \arctan x + 2 \int \frac{dx}{x\left(x^2 + 1\right)}. \text{ Also,}$$

$$\int \frac{dx}{x\left(x^2 + 1\right)} = \int \frac{\cos t}{\sin t}\, dt = \ln|\sin t| + C = \ln\left|\frac{x}{\sqrt{1 + x^2}}\right| + C.$$

$[x = \tan t,\, dx = \sec^2 t\, dt,\, \sin t = x/\sqrt{1 + x^2}]$.

Therefore, $\displaystyle\int \frac{x+2}{x\left(x^2+1\right)}\, dx = \arctan x + 2\ln\left|\frac{x}{\sqrt{1+x^2}}\right| + C$.

46. Using integration by parts, $\displaystyle\int x \arcsin x\, dx = \frac{1}{2}x^2 \arcsin x - \frac{1}{2}\int \frac{x^2}{\sqrt{1-x^2}}\, dx$. The integral on the right can be found using the substitution $x = \sin t$:

$\displaystyle\int \frac{x^2}{\sqrt{1-x^2}}\, dx = \frac{1}{2}\arcsin x - \frac{1}{2}x\sqrt{1-x^2} + C$. Therefore,

$$\int x \arcsin x\, dx = \frac{1}{2}x^2 \arcsin x - \frac{1}{4}\arcsin x + \frac{1}{4}x\sqrt{1-x^2} + C.$$

47. $\displaystyle\int \frac{\arctan x}{(1+x^2)^{3/2}}\, dx = \int w \cos w\, dw = w \sin w + \cos w = \frac{1 + x \arctan x}{\sqrt{1+x^2}} + C$.

48. $\displaystyle\int \frac{\arcsin x}{\left(1-x^2\right)^{3/2}}\, dx = \int t \sec^2 t\, dt = t \tan t - \int \tan t\, dt = t \tan t - \ln|\sec x|$

$$= \frac{x \arcsin x}{\sqrt{1-x^2}} + \ln(\sqrt{1-x^2}) + C.$$

Thus, $\displaystyle\int_{-1/2}^{1/2} \frac{\arcsin x}{\left(1-x^2\right)^{3/2}}\, dx = \left.\frac{x \arcsin x}{\sqrt{1-x^2}} + \ln(\sqrt{1-x^2})\right]_{-1/2}^{1/2} = 0$.

49. (a) Since $\cos(-v) = \cos v$ and $\sin(-v) = -\sin v$, $\cos(u-v) = \cos u \cos v + \sin u \sin v$. Therefore, $\cos(u+v) + \cos(u-v) = (\cos u \cos v - \sin u \sin v) + (\cos u \cos v + \sin u \sin v) = 2\cos u \cos v$.

(b) $\displaystyle\int \cos(ax)\cos(bx)\, dx = \frac{1}{2}\int \left(\cos((a+b)x) + \cos((a-b)x)\right) dx$

$$= \frac{1}{2(a+b)}\sin((a+b)x) + \frac{1}{2(a-b)}\sin((a-b)x) + C$$

50. (a) $\cos(u-v) - \cos(u+v) = (\cos u \cos v + \sin u \sin v) - (\cos u \cos v - \sin u \sin v) = 2\sin u \sin v$

(b) $\displaystyle\int \sin(ax)\sin(bx)\, dx = \frac{1}{2}\int \left(\cos((a-b)x) - \cos((a+b)x)\right) dx$

$$= \frac{1}{2(a-b)}\sin((a-b)x) - \frac{1}{2(a+b)}\sin((a+b)x) + C$$

51. (a) Since $\cos(-v) = \cos v$ and $\sin(-v) = -\sin v$, $\sin(u-v) = \sin u \cos v - \cos u \sin v$. Therefore,
$\sin(u+v) + \sin(u-v) = (\sin u \cos v + \cos u \sin v) + (\sin u \cos v - \cos u \sin v)$
$= 2\sin u \cos v$.

(b) $\displaystyle\int \sin(ax)\cos(bx)\,dx = \frac{1}{2}\int \Big(\sin\big((a+b)x\big) + \sin\big((a-b)x\big)\Big)\,dx$

$$= -\frac{1}{2(a+b)}\cos\big((a+b)x\big) - \frac{1}{2(a-b)}\cos\big((a-b)x\big) + C$$

52. (a) $\displaystyle\int \sec x\,dx = \int \frac{\cos x}{1 - \sin^2 x}\,dx = \int \frac{1}{1 - u^2}\,du$

$$= \frac{1}{2}\ln\left|\frac{1+u}{1-u}\right| = \frac{1}{2}\ln\left|\frac{1+\sin x}{1-\sin x}\right| + C.$$

(b) $\displaystyle\int \sec x\,dx = \frac{1}{2}\ln\left|\frac{1+\sin x}{1-\sin x}\right| = \frac{1}{2}\ln\left|\frac{1-\sin^2 x}{(1-\sin x)^2}\right| = \ln\left|\frac{\cos x}{1-\sin x}\right| + C.$

(c) $\displaystyle\int \sec x\,dx = \ln\left|\frac{\cos x}{1-\sin x}\right| = \ln\left|\frac{\cos x(1+\sin x)}{(1-\sin x)(1+\sin x)}\right|$

$$= \ln\left|\frac{\cos x + \cos x \sin x}{\cos^2 x}\right| + C$$

$$= \ln|\sec x + \tan x| + C.$$

53. (a) $u = \sin^{n-1} x \implies du = (n-1)\sin^{n-2} x \cos x\,dx;\; dv = \sin x\,dx \implies v = -\cos x.$ Thus,

$$\int \sin^n x\,dx = \int u\,dv = uv - \int v\,du = -\sin^{n-1} x \cos x + (n-1)\int \sin^{n-2} x \cos^2 x\,dx.$$

(b) $\displaystyle\int \sin^{n-2} x \cos^2 x\,dx = \int \sin^{n-2} x\,(1 - \sin^2 x)\,dx = \int \sin^{n-2} x\,dx - \int \sin^n x\,dx.$ Thus,

$$\int \sin^n x\,dx = -\sin^{n-1} x \cos x + (n-1)\int \sin^{n-2} x\,dx - (n-1)\int \sin^n x\,dx$$

so $\displaystyle n\int \sin^n x\,dx = -\sin^{n-1} x \cos x + (n-1)\int \sin^{n-2} x\,dx.$ Therefore, if $n \neq 0,$

$$\int \sin^n x\,dx = -\frac{\sin^{n-1} x \cos x}{n} + \frac{n-1}{n}\int \sin^{n-2} x\,dx.$$

54. Let $\displaystyle I_n = \int_0^{\pi/2} \sin^n x\,dx.$ Then, the reduction formula implies that

$$I_n = \int_0^{\pi/2} \sin^n x\,dx = -\frac{\sin^{n-1} x \cos x}{n}\Bigg]_0^{\pi/2} + \frac{n-1}{n}\int_0^{\pi/2} \sin^{n-2} x\,dx = \frac{n-1}{n}I_{n-2}.$$

Thus, since n is an odd integer and $I_1 = 1$, the formula

$$I_n = \frac{n-1}{n}\cdot\frac{n-3}{n-2}\cdots\frac{4}{5}\cdot\frac{2}{3}\cdot I_1 = \frac{n-1}{n}\cdot\frac{n-3}{n-2}\cdots\frac{4}{5}\cdot\frac{2}{3}$$

is valid for all odd integers $n \geq 3$.

55. Let $I_n = \int_0^{\pi/2} \sin^n x \, dx$. Then $I_1 = 1$, $I_2 = \pi/4$, and the reduction formula implies that $I_n = \frac{n-1}{n} I_{n-2}$.

    If n is an even integer such that $n > 2$,

    $$I_n = \frac{n-1}{n} I_{n-2} = \frac{n-1}{n} \frac{n-3}{n-2} I_{n-4} = \cdots = \frac{(n-1)}{n} \frac{(n-3)}{n-2} \cdots \frac{5}{6} \frac{3}{4} \frac{\pi}{4}.$$

56. Using integration by parts,

    $$\int \cos^n x \, dx = \int \cos^{n-1} x \cos x \, dx = \cos^{n-1} x \sin x + (n-1) \int \cos^{n-2} x \sin^2 x \, dx$$

    $$= \cos^{n-1} x \sin x + (n-1) \int \cos^{n-2} x \, dx - (n-1) \int \cos^n x \, dx.$$

    Thus, $n \int \cos^n x \, dx = \cos^{n-1} x \sin x + (n-1) \int \cos^{n-2} x \, dx$ so

    $$\int \cos^n x \, dx = \frac{\cos^{n-1} x \sin x}{n} + \frac{n-1}{n} \int \cos^{n-2} x \, dx.$$

57. $$\int \tan^n x \, dx = \int \tan^{n-2} x \tan^2 x \, dx = \int \tan^{n-2} x (\sec^2 x - 1) \, dx$$

    $$= \int \tan^{n-2} x \sec^2 x \, dx - \int \tan^{n-2} x \, dx = \frac{\tan^{n-1} x}{n-1} - \int \tan^{n-2} x \, dx.$$

58. Using integration by parts,

    $$\int \sec^n x \, dx = \int \sec^{n-2} x \sec^2 x \, dx = \sec^{n-2} x \tan x - (n-2) \int \sec^{n-3} x \sec x \tan^2 x \, dx$$

    $$= \sec^{n-2} x \tan x - (n-2) \int \sec^{n-2} x \tan^2 x \, dx$$

    $$= \sec^{n-2} x \tan x + (n-2) \int \sec^{n-2} x \, dx - (n-2) \int \sec^n x \, dx.$$

    Thus, $(n-1) \int \sec^n x \, dx = \sec^{n-2} x \tan x + (n-2) \int \sec^{n-2} x \, dx$ so

    $$\int \sec^n x \, dx = \frac{\sec^{n-2} x \tan x}{n-1} + \frac{n-2}{n-1} \int \sec^{n-2} x \, dx.$$

59. Draw a right triangle with hypotenuse $\sqrt{1 + t^2}$ and sides of length 1 and t.

    (a) If the angle opposite the side of length t is $x/2$, then $\sin(x/2) = t/\sqrt{1 + t^2}$ since $\sin \theta = $ opposite/hypotenuse.

(b) If the angle opposite the side of length t is $x/2$, then $\cos(x/2) = 1/\sqrt{1+t^2}$ since $\cos\theta = $ adjacent/hypotenuse.

(c) The desired result follows from the identity $\sin(2\theta) = 2\cos\theta\sin\theta$.

(d) The desired result follows from the identity $\cos(2\theta) = 2\cos^2\theta - 1$.

60. Let $x = 2\arctan t$. Then $dx = \dfrac{2}{1+t^2}\,dt$ and

$$\int \frac{dx}{1+\sin x + \cos x} = \int \frac{1}{1 + \frac{2t}{1+t^2} + \frac{1-t^2}{1+t^2}} \cdot \frac{2}{1+t^2}\,dt$$

$$= \int \frac{dt}{1+t} = \ln|1+t| + C = \ln\left|1 + \tan(x/2)\right| + C.$$

61. $\displaystyle\int \frac{dx}{1+\cos x} = \int \frac{1}{1 + (1-t^2)/(1+t^2)} \frac{2}{1+t^2}\,dt = \int dt = t + C = \tan(x/2) + C.$

62. Using the substitution $x = 2\arctan t$,

$$\int_0^{\pi/2} \frac{dx}{2\cos x + 3\sin x} = \int_0^1 \frac{dt}{1 + 3t - t^2} = -\int_0^1 \frac{dt}{(t-3/2)^2 - 13/4}$$

$$= -\frac{1}{\sqrt{13}} \ln\left|\frac{(t-3/2) - \sqrt{13}/2}{(t-3/2) + \sqrt{13}/2}\right|\Bigg]_0^1$$

$$= \frac{1}{\sqrt{13}} \ln\left|\frac{-3/2 - \sqrt{13}/2}{-3/2 + \sqrt{13}/2}\right| - \frac{1}{\sqrt{13}} \ln\left|\frac{-1/2 - \sqrt{13}/2}{-1/2 + \sqrt{13}/2}\right|.$$

63. Using the substitution $u = 2 + \cos x$, $\displaystyle\int \frac{\sin x}{2+\cos x}\,dx = -\int \frac{du}{u} = -\ln(2+\cos x).$ Thus,

$$\int_{\pi/4}^{\pi} \frac{\sin x}{2+\cos x}\,dx = -\ln(2+\cos x)\Bigg]_{\pi/4}^{\pi} = \ln(2 + \sqrt{2}/2) = \ln(4 + \sqrt{2}) - \ln 2.$$

64. The substitution transforms $\sin x$ and $\cos x$ into rational expressions of t. Since any trigonometric function can be expressed in terms of $\sin x$ and $\cos x$, the substitution transforms any rational expression involving only constants and powers of trigonometric functions into a rational expression in t. (The composition of rational functions is a rational function.)

65. (a) Let $u = 1/(1+x^2)^n$ and $dv = dx$. Then, $du = -2nx/(1+x^2)^{n+1}\,dy$ and $v = x$ so

$$I_n = \int \frac{dx}{(1+x^2)^n} = \frac{x}{(1+x^2)^n} + 2n\int \frac{x^2}{(1+x^2)^{n+1}}\,dx$$

$$= \frac{x}{(1+x^2)^n} + 2n\int \frac{1+x^2}{(1+x^2)^{n+1}}\,dx - 2n\int \frac{dx}{(1+x^2)^{n+1}}$$

$$= \frac{x}{(1+x^2)^n} + 2n\int \frac{dx}{(1+x^2)^n} - 2n\int \frac{dx}{(1+x^2)^{n+1}}.$$

Thus,

$$I_{n+1} = \frac{x}{2n(1+x^2)^n} - \frac{1-2n}{2n} I_n$$

or, equivalently,

$$I_n = \frac{x}{2(n-1)(1+x^2)^{n-1}} + \frac{2n-3}{2n-2} I_{n-1}.$$

(b) If $x = \tan u$, then $dx = \sec^2 u\, du$ and $I_n = \displaystyle\int \frac{dx}{(1+x^2)^n} = \int \cos^{2n-2} u\, du$. Now, using the reduction formula for powers of the cosine function,

$$
\begin{aligned}
I_n = \int \frac{dx}{(1+x^2)^n} &= \int \cos^{2n-2} u\, du \\
&= \frac{\cos^{2n-3} u \sin u}{2n-2} + \frac{2n-3}{2n-2} \int \cos^{2n-4} u\, du \\
&= \frac{x}{(2n-2)(1+x^2)^{n-1}} + \frac{2n-3}{2n-2} I_{n-1}.
\end{aligned}
$$

§8.4   Miscellaneous Antiderivatives

1. $\displaystyle\int \frac{\sin x}{(3 + \cos x)^2}\, dx = \frac{1}{3 + \cos x} + C.$
   [substitution: $u = 3 + \cos x$.]

2. $\displaystyle\int \frac{x^2}{x + 1}\, dx = \frac{1}{2}(x + 1)^2 - 2(x + 1) + \ln|x + 1| + C.$
   [substitution: $u = x + 1$.]

3. $\displaystyle\int x\left(3 + 4x^2\right)^5 dx = \frac{1}{48}\left(3 + 4x^2\right)^6 + C.$
   [substitution: $u = 3 + 4x^2$.]

4. $\displaystyle\int \frac{dx}{\sqrt{1 - x^2}} = \arcsin x + C.$

5. $\displaystyle\int \frac{x}{\sqrt[3]{x^2 + 4}}\, dx = \frac{3}{4}\left(x^2 + 4\right)^{2/3} + C.$
   [substitution: $u = x^2 + 4$.]

6. $\displaystyle\int \frac{dx}{x(3x - 2)} = \int \left(\frac{3}{2(3x - 2)} - \frac{1}{2x}\right) dx = \left(\ln|3x - 2| - \ln|x|\right)/2 + C.$

7. $\displaystyle\int \frac{(\ln x)^2}{x}\, dx = \frac{1}{3}\left(\ln|x|\right)^3 + C.$
   [substitution: $u = \ln x$.]

8. $\displaystyle\int e^x \sin x\, dx = \frac{1}{2}e^x\left(\sin x - \cos x\right) + C.$
   [integration by parts (twice).]

9. $\displaystyle\int \frac{\ln x}{x}\, dx = \frac{1}{2}\left(\ln|x|\right)^2 + C.$
   [substitution: $u = \ln x$] .

10. $\displaystyle\int x\sqrt{x + 2}\, dx = \frac{2}{5}(x + 2)^{5/2} - \frac{4}{3}(x + 2)^{3/2} + C.$
    [substitution: $u = x + 2$.]

11. $\displaystyle\int \frac{x}{3x + 2}\, dx = \frac{1}{3}\int \left(1 - \frac{2}{3x + 2}\right) dx = \frac{x}{3} - \frac{2}{9}\ln|3x + 2| + C.$

12. Using integration by parts (with $u = x$ and $dv = \cos x\, dx$)
    $$\int x \cos x\, dx = x \sin x + \cos x + C.$$

13. $\displaystyle\int \sin^2(3x)\cos(3x)\,dx = \frac{1}{9}\sin^3(3x) + C.$

[substitution: $u = \sin(3x)$.]

14. $\displaystyle\int xe^{3x}\,dx = \frac{1}{3}e^{3x}\left(x - \frac{1}{3}\right) + C.$

[integration by parts: $u = x,\, dv = e^{3x}\,dx$.]

15. $\displaystyle\int xe^{3x^2}\,dx = \frac{1}{6}e^{3x^2} + C.$

[substitution: $u = 3x^2$.]

16. $\displaystyle\int \frac{dx}{1 + 4x^2} = \frac{1}{2}\arctan(2x) + C.$

[substitution: $u = 2x$.]

17. $\displaystyle\int (2 - 3x)^{10}\,dx = -\frac{1}{33}(2 - 3x)^{11} + C.$

[substitution: $u = 2 - 3x$.]

18. $\displaystyle\int \arctan x\,dx = x\arctan x - \frac{1}{2}\ln(1 + x^2) + C.$

[integration by parts: $u = \arctan x,\, dv = dx$.]

19. $\displaystyle\int \frac{\sec^2 x}{3 + \tan x}\,dx = \ln|3 + \tan x| + C.$

[substitution: $u = 3 + \tan x$.]

20. $\displaystyle\int x\sin x\,dx = \sin x - x\cos x + C.$

[integration by parts: $u = x,\, dv = \sin x\,dx$.]

21. $\displaystyle\int \frac{dx}{(x - 1)(x + 2)} = \frac{1}{3}\ln|x - 1| - \frac{1}{3}\ln|x + 2| = \frac{1}{3}\ln\left|\frac{x - 1}{x + 2}\right| + C.$

[partial fractions: $\frac{1}{(x-1)(x+2)} = \frac{1}{3(x-1)} - \frac{1}{3(x+2)}$.]

22. $\displaystyle\int x^2 \ln x\,dx = \frac{1}{3}x^3\ln|x| - \frac{1}{9}x^3 + C.$

[integration by parts: $u = \ln x,\, dv = x^2\,dx$.]

23. $\displaystyle\int \frac{2x + 3}{4x + 5}\,dx = \frac{1}{2}\int\left(1 + \frac{1}{4x + 5}\right)dx = \frac{x}{2} + \frac{\ln|4x + 5|}{8} + C.$

24. $\displaystyle\int \frac{x + 1}{x^2 + 1}\,dx = \int \frac{x}{x^2 + 1}\,dx + \int \frac{1}{x^2 + 1}\,dx = \frac{1}{2}\ln(x^2 + 1) + \arctan x + C.$

25. $\int \dfrac{e^x}{\sqrt{1-e^{2x}}}\,dx = \arcsin\left(e^x\right) + C.$

    [substitution: $u = e^x$.]

26. $\int \dfrac{\sin x}{2 + \cos x}\,dx = -\ln(2 + \cos x) + C.$

    [substitution: $u = 2 + \cos x$.]

27. $\int \ln x\,dx = x(\ln x - 1) + C.$

    [integration by parts: $u = \ln x,\, dv = dx$.]

28. $\int x \cos\left(3x^2\right)\,dx = \dfrac{1}{6}\sin(3x^2) + C.$

    [substitution: $u = 3x^2$.]

29. $\int \arcsin x\,dx = x \arcsin x + \sqrt{1 - x^2} + C.$

    [integration by parts: $u = \arcsin x,\, dv = dx$.]

30. Let $u = \sqrt{x}$. Then,

$$\int \frac{\sqrt{x}}{1 + x}\,dx = 2 \int \frac{u^2}{1 + u^2}\,du = 2 \int \left(1 - \frac{1}{1 + u^2}\right)\,du$$
$$= 2u - 2\arctan u = 2\sqrt{x} - 2\arctan(\sqrt{x}) + C.$$

31. $\int \dfrac{dx}{x^2 + 2x + 3} = \dfrac{1}{\sqrt{2}}\arctan\left(\dfrac{x+1}{\sqrt{2}}\right) + C.$

    [complete the square, substitution: $x^2 + 2x + 3 = (x + 1)^2 + 2 = u^2 + 2$.]

32. $\int \dfrac{x}{\sqrt{x-2}}\,dx = \dfrac{2}{3}(x - 2)^{3/2} + 4\sqrt{x - 2} + C.$

    [substitution: $u = x - 2$.]

33. $\int \dfrac{dx}{\sqrt{1 - 4x^2}} = \dfrac{1}{2}\arcsin(2x) + C.$

    [substitution: $u = 2x$.]

34. $\int \dfrac{x^3}{1 + x^2}\,dx = \dfrac{1}{2}(1 + x^2) - \dfrac{1}{2}\ln(1 + x^2) + C.$

    [substitution: $u = 1 + x^2$.]

35. $\int \tan x\,dx = -\ln|\cos x| + C.$

    [Write $\tan x = \sin x / \cos x$, then use the substitution $u = \cos x$.]

36. $\displaystyle\int \cos(2x)\,dx = \frac{1}{2}\sin(2x) + C.$

37. $\displaystyle\int e^{2x}\sqrt{1+e^x}\,dx = \frac{2}{5}\left(1+e^x\right)^{5/2} - \frac{2}{3}\left(1+e^x\right)^{3/2} + C.$

    [substitution: $u = 1 + e^x$.]

38. $\displaystyle\int \frac{dx}{1+x^2} = \arctan x + C.$

39. Let $u = (x+1)/2$. Then, $du = \frac{1}{2}dx$ and

    $$\int \frac{dx}{\sqrt{3 - 2x - x^2}} = \int \frac{du}{\sqrt{1 - u^2}} = \arcsin u = \arcsin\left(\tfrac{1}{2}(x+1)\right) + C.$$

40. Using integration by parts with $u = \arcsin x$ and $dv = x^2\,dx$,

    $$\int x^2 \arcsin x\,dx = \frac{1}{3}x^3 \arcsin x - \frac{1}{3}\int \frac{x^3}{\sqrt{1-x^2}}\,dx.$$

    To find the remaining antiderivative, use the substitution $w = 1 - x^2$:

    $$\int \frac{x^3}{\sqrt{1-x^2}}\,dx = -\frac{1}{2}\int \frac{1-w}{\sqrt{w}}\,dw = -\frac{1}{2}\int \frac{dw}{\sqrt{w}} + \frac{1}{2}\int \sqrt{w}\,dw$$

    $$= -\sqrt{w} + \frac{1}{3}w^{3/2} + C = -\sqrt{1-x^2} + \frac{1}{3}(1-x^2)^{3/2} + C.$$

    Therefore, $\displaystyle\int x^2 \arcsin x\,dx = \frac{1}{3}x^3 \arcsin x + \frac{1}{3}\sqrt{1-x^2} - \frac{1}{9}(1-x^2)^{3/2} + C.$

41. Let $u = \ln x$. Then, $du = dx/x$ and $\displaystyle\int \frac{dx}{x(\ln x)^2} = \int \frac{du}{u^2} = -\frac{1}{u} + C = -\frac{1}{\ln|x|} + C.$

42. $\displaystyle\int x \arctan x\,dx = \frac{1}{2}(x^2+1)\arctan x - \frac{x}{2} + C.$

    [integration by parts: $u = \arctan x$, $dv = x\,dx$.]

43. $\displaystyle\int \frac{dx}{9x^2 - 4} = \int \frac{dx}{(3x-2)(3x+2)} = \frac{1}{4}\int \left(\frac{1}{3x-2} - \frac{1}{3x+2}\right)dx$

    $= \left(\ln|3x-2| - \ln|3x+2|\right)/12 + C.$

44. $\displaystyle\int \frac{x+5}{x^2+3x-4}\,dx = \frac{6}{5}\ln|x-1| - \frac{1}{5}\ln|x+4| + C.$

    [partial fractions: $\frac{x+5}{x^2+3x-4} = \frac{6}{5(x-1)} - \frac{1}{5(x+4)}$.]

45. $\displaystyle\int \frac{x^3}{\sqrt{4-x^2}}\,dx = -4\sqrt{4-x^2} + \frac{1}{3}\left(4-x^2\right)^{3/2} + C.$

    [substitution ($u = 4 - x^2$) or integration by parts ($u = x^2$, $dv = x/\sqrt{4-x^2}\,dx$).]

46. $\displaystyle\int \frac{dx}{\sqrt[3]{x-1}} = \frac{3}{2}(x-1)^{2/3} + C.$

47. $\displaystyle\int \frac{x}{(x-1)(x+1)}\, dx = \frac{1}{2}\ln|x+1| + \frac{1}{2}\ln|x-1| = \frac{1}{2}\ln\left|x^2-1\right| + C.$

    [partial fractions: $\frac{x}{(x-1)(x+1)} = \frac{1}{2(x+1)} + \frac{1}{2(x-1)}$.]

48. $\displaystyle\int x^3 e^{x^2}\, dx = \frac{1}{2}e^{x^2}(x^2-1) + C.$

    [substitution ($w = x^2$), then integration by parts ($u = w$, $dv = e^w\, dw$).]

49. $\displaystyle\int \frac{dx}{\sqrt{9+x^2}} = \ln\left|x + \sqrt{9+x^2}\right| + C.$

    [trigonometric substitution: $x = 3\tan t$.]

50. $\displaystyle\int \frac{dx}{2x-x^2} = \frac{1}{2}\ln|x| - \frac{1}{2}\ln|x-2| = \frac{1}{2}\ln\left|\frac{x}{x-2}\right| + C.$

    [partial fractions: $\frac{1}{2x-x^2} = \frac{1}{x(2-x)} = \frac{1}{2x} + \frac{1}{2(2-x)}$.]

51. $\displaystyle\int \frac{x^2}{1-3x}\, dx = -\frac{1}{54}(1-3x)^2 + \frac{2}{27}(1-3x) - \frac{1}{27}\ln|1-3x| + C$

    $\displaystyle\qquad\qquad = -\frac{1}{6}x^2 - \frac{1}{9}x - \frac{1}{27}\ln|1-3x| + C.$

    [substitution ($u = 1 - 3x$) or partial fractions.]

52. $\displaystyle\int \frac{x}{\left(x^2-1\right)^3}\, dx = -\frac{1}{4\left(x^2-1\right)^2} + C.$

    [substitution: $u = x^2 - 1$.]

53. $\displaystyle\int e^x e^{2x}\, dx = \int e^{3x}\, dx = \frac{1}{3}e^{3x} + C.$

54. $\displaystyle\int \sqrt{4x-3}\, dx = \frac{1}{6}(4x-3)^{3/2} + C.$

    [substitution: $u = 4x - 3$.]

55. $\displaystyle\int \ln(1+x^2)\, dx = x\ln(1+x^2) - 2x + 2\arctan x + C.$

    [integration by parts: $u = \ln(1+x^2)$, $dv = dx$.]

56. $\displaystyle\int \sin(\sqrt{x})\, dx = 2\sin(\sqrt{x}) - 2\sqrt{x}\cos(\sqrt{x}) + C.$

    [substitution ($w = \sqrt{x}$), then integration by parts ($u = w$, $dv = \sin w\, dw$.]

57. Using integration by parts $\int x \arcsin x \, dx = \dfrac{x^2 \arcsin x}{2} - \dfrac{1}{2} \int \dfrac{x^2}{\sqrt{1-x^2}} \, dx$. Now, using the trigonometric substitution $x = \sin t$,

$$\int \frac{x^2}{\sqrt{1-x^2}} \, dx = \int \sin^2 t \, dt = (-\sin t \cos t + t)/2$$
$$= \left(-x\sqrt{1-x^2} + \arcsin x\right)/2.$$

Thus,

$$\int x \arcsin x \, dx = \frac{x^2 \arcsin x}{2} + \frac{x\sqrt{1-x^2}}{4} - \frac{\arcsin x}{4} + C.$$

58. $\int \dfrac{dx}{9-x^2} = \dfrac{1}{6} \int \left(\dfrac{1}{3+x} + \dfrac{1}{3-x} \right) dx = (\ln|x+3| - \ln|3-x|)/6.$

59. Let $u = 2x + 3$. Then, $\int \dfrac{dx}{\sqrt{2x+3}} = \dfrac{1}{2} \int \dfrac{du}{\sqrt{u}} = \sqrt{u} = \sqrt{2x+3} + C.$

60. $\int \dfrac{dx}{\left(4-x^2\right)^{3/2}} = \dfrac{x}{4\sqrt{4-x^2}} + C.$

[trigonometric substitution: $x = 2\sin t$.]

61. Let $u = 2x + 3$. Then,

$$\int \frac{x}{(2x+3)^4} \, dx = \frac{1}{4} \int \frac{u-3}{u^4} \, du = -\frac{1}{8u^2} + \frac{1}{4u^3}$$
$$= -\frac{1}{8(2x+3)^2} + \frac{1}{4(2x+3)^3} = -\frac{2x+1}{8(2x+3)^3} + C.$$

62. $\int x\sqrt{2x+1} \, dx = \dfrac{1}{10}(2x+1)^{5/2} - \dfrac{1}{6}(2x+1)^{3/2} + C.$

[substitution: $u = 2x + 1$.]

63. $\int \dfrac{\tan x}{\sec^2 x} \, dx = -\dfrac{1}{2}\cos^2 x + C.$

[Write $\frac{\tan x}{\sec^2 x} = \sin x \cos x$, then use substitution ($u = \cos x$).]

64. $\int \dfrac{x}{16+9x^2} \, dx = \dfrac{1}{18} \ln\left(16+9x^2\right) + C.$

[subsitution: $u = 16 + 9x^2$.]

65. $\int \dfrac{dx}{e^x - 1} = \ln\left|1 - e^{-x}\right| + C.$

[Write $\frac{1}{e^x-1} = \frac{e^{-x}}{1-e^{-x}}$, then use subsitution $u = 1 - e^{-x}$.]

66. $\int \dfrac{dx}{\sqrt{2x - x^2}} = -\arcsin(1 - x) + C.$

    [Write $2x - x^2 = 1 - (1 - x)^2$, then substitute $u = 1 - x$.]

67. $\int \dfrac{dx}{1 + \sqrt{x}} = 2(1 + \sqrt{x}) - 2\ln(1 + \sqrt{x}) + C.$

    [subsitution: $u = 1 + \sqrt{x}$.]

68. $\int \dfrac{x^3\, dx}{\left(x^2 + 1\right)^2} = \dfrac{1}{2(x^2 + 1)} + \dfrac{1}{2}\ln(x^2 + 1) + C.$

    [substitution $u = x^2 + 1$.]

69. $\int x^2 \ln(3x)\, dx = \dfrac{1}{3}x^3 \ln(3x) - \dfrac{1}{9}x^3 + C.$

    [integration by parts: $u = \ln(3x),\, dv = x^2\, dx$.]

70. $\int \dfrac{x}{9 + 4x^4}\, dx = \dfrac{1}{12}\arctan\left(\dfrac{2x^2}{3}\right) + C.$

71. $\int \sqrt{x}\, \ln x\, dx = \dfrac{2}{3}x^{3/2}\ln|x| - \dfrac{4}{9}x^{3/2} + C.$

    [integration by parts: $u = \ln x,\, dv = \sqrt{x}\, dx$.]

72. $\int x\sec^2 x\, dx = x\tan x + \ln|\cos x| + C.$

    [integration by parts: $u = x,\, dv = \sec^2 x\, dx$.]

73. $\int \dfrac{7 - x}{(x + 3)(x^2 + 1)}\, dx = \ln|x + 3| + 2\arctan x - \dfrac{1}{2}\ln(x^2 + 1) + C.$

    [partial fractions: $\frac{7-x}{(x+3)(x^2+1)} = \frac{1}{x+3} + \frac{2-x}{x^2+1}$.]

74. $\int \dfrac{x + 6}{(x + 1)\left(x^2 + 4\right)}\, dx = \ln|x + 1| + \arctan(x/2) - \dfrac{1}{2}\ln(x^2 + 4) + C.$

    [partial fractions: $\frac{x+6}{(x+1)(x^2+4)} = \frac{1}{x+1} + \frac{2-x}{x^2+4}$.]

75. $\int x\sin^2 x\cos x\, dx = \dfrac{1}{3}x\sin^3 x + \dfrac{1}{3}\cos x - \dfrac{1}{9}\cos^3 x$

    $\qquad\qquad\qquad = \dfrac{1}{3}x\sin^3 x + \dfrac{1}{9}\sin^2 x\cos x + \dfrac{2}{9}\cos x + C.$

    [integration by parts with $u = x$ and $dv = \sin^2 x\cos x\, dx$.]

76. Let $u = \sin x$. Then, $du = \cos x\, dx$ and

$$\int \sin^3 x \cos^3 x\, dx = \int \sin^3 x \cos^2 x \cos x\, dx = \int \sin^3 x (1 - \sin^2 x) \cos x\, dx$$
$$= \int \sin^3 x \cos x\, dx - \int \sin^5 x \cos x\, dx = \int u^3\, du - \int u^5\, du$$
$$= \tfrac{1}{4} u^4 - \tfrac{1}{6} u^6 + C = \tfrac{1}{4} \sin^4 x - \tfrac{1}{6} \sin^6 x + C.$$

77. $\displaystyle \int \frac{dx}{x^3 + x} = \ln |x| - \frac{1}{2} \ln(x^2 + 1) + C.$

[partial fractions: $\frac{1}{x^3+x} = \frac{1}{x} - \frac{x}{x^2+1}$.]

78. $\displaystyle \int \tan^4 x\, dx = \frac{1}{3} \tan^3 x - \tan x + x + C.$

79. $\displaystyle \int \left(x^2 + 2x + 3 \right)^{3/2} dx = \frac{1}{4}(x + 1)(x^2 + 2x + 3)^{3/2} + \frac{3}{4}(x + 1)\sqrt{x^2 + 2x + 3} +$

$\displaystyle \frac{3}{2} \ln \left| \frac{\sqrt{x^2 + 2x + 3}}{\sqrt{2}} + \frac{x + 1}{\sqrt{2}} \right| + C.$

[Write $x^2 + 2x + 3 = (x + 1)^2 + 2$, then use a trigonometric substitution $(x + 1 = \sqrt{2} \tan t)$.]

80. $\displaystyle \int \sin(3x) \cos(5x)\, dx = \frac{1}{4} \cos(2x) - \frac{1}{16} \cos(8x) + C.$

[Write $\sin(3x) \cos(5x) = \frac{1}{2} \sin(8x) - \frac{1}{2} \sin(2x)$.]

81. Let $u = \sqrt{1 + e^x}$. Then,

$$\int \sqrt{1 + e^x}\, dx = 2 \int \frac{u^2}{u^2 - 1}\, du = 2 \int \left(1 + \frac{1}{u^2 - 1} \right) du$$
$$= 2 \int \left(1 - \frac{1}{2(u + 1)} + \frac{1}{2(u - 1)} \right) du$$
$$= 2u - \ln |u + 1| + \ln |u - 1|$$
$$= 2\sqrt{1 + e^x} - \ln \left| \sqrt{1 + e^x} + 1 \right| + \ln \left| \sqrt{1 + e^x} - 1 \right| + C.$$

82. $\displaystyle \int \frac{dx}{x \left(x + \sqrt[3]{x} \right)} = -\frac{3}{\sqrt[3]{x}} - 3 \arctan(\sqrt[3]{x}) + C.$

[substitution $(u = x^{1/3})$, then partial fractions $(\frac{1}{u^4 + u^2} = \frac{1}{u^2} - \frac{1}{u^2 + 1})$.]

83. $\displaystyle \int \frac{dx}{x^3 + 1} = \frac{1}{3} \ln |x + 1| - \frac{1}{6} \ln |x^2 - x + 1| + \frac{1}{\sqrt{3}} \arctan \left(\frac{2x - 1}{\sqrt{3}} \right) + C.$

[Write $x^3 + 1 = (x + 1)(x^2 - x + 1) = (x + 1) \cdot \left((2x - 1)^2 + 3 \right) / 4$, then use partial fractions, etc.]

84. $\displaystyle\int \frac{dx}{(e^x - e^{-x})^2} = \frac{1}{4}\left(\frac{1}{1+e^x} + \frac{1}{1-e^x}\right) = \frac{1}{2 - 2e^{2x}} + C.$

[Write $\left(e^x - e^{-x}\right)^{-2} = e^{2x}\left(e^{2x} - 1\right)^{-2} = (e^x)^2\left(e^x - 1\right)^{-2}\left(e^x + 1\right)^{-2}$, then use substitution $(u = e^x)$ and partial fractions.]

85. $\displaystyle\int x \tan^2 x \, dx = x \tan x - \frac{1}{2}x^2 - \ln|\sec x| + C.$

[integration by parts: $u = x$, $dv = \tan^2 x \, dx$.]

86. $\displaystyle\int \cos^3 x \, dx = \sin x - \frac{1}{3}\sin^3 x + C.$

[Write $\cos^3 x = \cos x \cos^2 x = \cos x - \cos x \sin^2 x$, then use the substitution $u = \sin x$.]

87. $\displaystyle\int \sin x \sin(2x) \, dx = \frac{1}{2}\sin x - \frac{1}{6}\sin(3x) + C.$

[Write $\sin x \sin(2x) = \frac{1}{2}\cos x - \frac{1}{2}\cos(3x)$.]

88. $\displaystyle\int \sin^5 x \cos^2 x \, dx = \frac{1}{7}\sin^6 x \cos x - \frac{1}{35}\sin^4 x \cos x - \frac{4}{105}\sin^2 x \cos x - \frac{8}{105}\cos x + C.$

[Write $\cos^2 x = 1 - \sin^2 x$, then use a reduction formula.]

9   *Function Approximation*

§9.1   Taylor Polynomials

1. (a) If f is a polynomial of degree n, the Maclaurin series for f is just f itself:
   $1 + 2x + 44x^2 - 12x^3 + x^4$.

   (b) Finding values and derivatives at $x = 3$ gives
   $f(x) = 160 + 50(x - 3) - 10(x - 3)^2 + (x - 3)^4$.

2. Let $f(x) = x^4 - 4x^3 + 5x$; we want the Taylor series expansion at $x = 1$. Because
   $f(1) = 2$, $f'(1) = -3$, $f''(1) = -12$, $f'''(1) = 0$, $f^{(4)}(1) = 24$, and $f^{(k)}(1) = 0$ for all
   $k \geq 5$, Taylor's theorem gives $f(x) = 2 - 3(x - 1) - 6(x - 1)^2 + (x - 1)^4$. Thus, $a_0 = 2$,
   $a_1 = -3$, $a_2 = -6$, $a_3 = 0$, and $a_4 = 1$.

3. Since p is a polynomial, it's easiest to find $p(1)$ and $p^{(4)}(1)$ from the coefficients of p:
   $p(1) = 9$ and $p^{(4)}(1)/24 = 1/5$.

   (a) $p(1) = 9$

   (b) $p^{(4)}(1) = 24/5$

4. Since q is a polynomial, it's easiest to find $q(2)$ and $q^{(9)}(2)$ from the coefficients of q:
   $q''(2)/2 = 1/4$ and $q^{(4)}(2)/9! = 1/81$.

   (a) $q''(2) = 1/2$

   (b) $q^{(9)}(2) = 4480$

5. Let a_k be the coefficient of $(x + 3)^k$. Then $a_k = p^{(k)}(-3)/k!$. In particular:

   (a) $p^{(11)}(-3) = a_{11} \times 11! = 0$.

   (b) $p^{(12)}(-3) = a_{12} \times 12! = \dfrac{1}{13^4} \times 12! \approx 16771$.

6. Let a_k be the coefficient of $(x - 3)^k$ in $q(x)$. Then $a_k = q^{(k)}(3)/k!$. In particular:

   (a) $q^{(99)}(3) = a_{99} \times 99! = \dfrac{99!}{49!}$.

   (b) $q^{(100)}(3) = a_{100} \times 100! = 0$.

7. Finding values and derivatives at $x = 0$ gives $P_6(x) = 1 - \dfrac{x^2}{2} + \dfrac{x^4}{24} - \dfrac{x^6}{720}$; plotting f and
   P_6 together near $x = 0$ shows that the approximation error is less than 0.01 on $[-2.13, 2.13]$
   (or on any smaller interval centered at zero).

8. Finding values and derivatives at $x = 0$ gives
$$P_6(x) = 1 + x + \frac{x^2}{2} + \frac{x^3}{6} + \frac{x^4}{24} + \frac{x^5}{120} + \frac{x^6}{720};$$ plotting f and P_6 together near $x = 0$ shows that the approximation error is less than 0.01 on $[-1.69, 1.69]$ (or on any smaller interval centered at zero).

9. Finding values and derivatives at $x = 0$ gives $P_6(x) = x - \frac{x^2}{2} + \frac{x^3}{3} - \frac{x^4}{4} + \frac{x^5}{5} - \frac{x^6}{6}$; plotting f and P_6 together near $x = 0$ shows that the approximation error is less than 0.01 on $[-.61, .73]$ (or on any smaller interval centered at zero).

10. Finding values and derivatives at $x = 0$ gives
$$P_6(x) = 1 + \frac{x}{2} - \frac{x^2}{8} + \frac{x^3}{16} - \frac{5x^4}{128} + \frac{7x^5}{256} - \frac{21x^6}{1024};$$ plotting f and P_6 together near $x = 0$ shows that the approximation error is less than 0.01 on $[-0.80, 1.02]$ (or on any smaller interval centered at zero).

11. Finding values and derivatives at $x = 0$ gives $P_6(x) = x + \frac{x^3}{6} + \frac{x^5}{120}$; plotting f and P_6 together near $x = 0$ shows that the approximation error is less than 0.01 on $[-1.74, 1.74]$ (or on any smaller interval centered at zero).

12. Finding values and derivatives at $x = 0$ gives $P_6(x) = 1 + \frac{x^2}{2} + \frac{x^4}{24} + \frac{x^6}{720}$; plotting f and P_6 together near $x = 0$ shows that the approximation error is less than 0.01 on $[-2.10, 2.10]$ (or on any smaller interval centered at zero).

13. Finding values and derivatives at $x = \pi/4$ gives
$$P_5(x) = \frac{1}{\sqrt{2}} \left(1 + x - \pi/4 - \frac{(x - \pi/4)^2}{2} - \frac{(x - \pi/4)^3}{6} + \frac{(x - \pi/4)^4}{24} + \frac{(x - \pi/4)^5}{120} \right);$$ plotting f and P_5 together near $x = \pi/4$ shows that the approximation error is less than 0.01 on $[-0.76, 2.22]$ (or on any smaller interval centered at $\pi/4$).

14. Finding values and derivatives at $x = \pi/4$ gives
$$P_5(x) = \frac{1}{\sqrt{2}} \left(1 - x - \pi/4 - \frac{(x - \pi/4)^2}{2} + \frac{(x - \pi/4)^3}{6} + \frac{(x - \pi/4)^4}{24} - \frac{(x - \pi/4)^5}{120} \right);$$ plotting f and P_5 together near $x = \pi/4$ shows that the approximation error is less than 0.01 on $[-0.65, 2.33]$ (or on any smaller interval centered at $\pi/4$).

15. Finding values and derivatives at $x = \pi/3$ gives

$$P_5(x) = \frac{\sqrt{3}}{2} + \frac{x - \pi/3}{2} - \frac{\sqrt{3}}{4} (x - \pi/3)^2 - \frac{1}{12} (x - \pi/3)^3$$
$$+ \frac{1}{16\sqrt{3}} (x - \pi/3)^4 + \frac{1}{240} (x - \pi/3)^5;$$

plotting f and P_5 together near $x = \pi/3$ shows that the approximation error is less than 0.01 on $[-0.41, 2.45]$ (or on any smaller interval centered at $\pi/3$).

16. Finding values and derivatives at $x = \pi/3$ gives

$$P_5(x) = \frac{1}{2} - \frac{\sqrt{3}}{2}(x - \pi/3) - \frac{1}{4}(x - \pi/3)^2 + \frac{1}{4\sqrt{3}}(x - \pi/3)^3$$
$$+ \frac{1}{48}(x - \pi/3)^4 - \frac{1}{80\sqrt{3}}(x - \pi/3)^5;$$

plotting f and P_5 together near $x = \pi/3$ shows that the approximation error is less than 0.01 on $[-0.44, 2.78]$ (or on any smaller interval centered at $\pi/3$).

17. Finding values and derivatives at $x = 1$ gives
$P_5(x) = 1 - (x - 1) + (x - 1)^2 - (x - 1)^3 + (x - 1)^4 - (x - 1)^5$; plotting f and P_5 together near $x = 1$ shows that the approximation error is less than 0.01 on $[0.58, 1.49]$ (or on any smaller interval centered at $x = 1$).

18. Finding values and derivatives at $x = 2$ gives
$P_5(x) = \frac{1}{2} - \frac{(x - 2)}{4} + \frac{(x - 2)^2}{8} - \frac{(x - 2)^3}{16} + \frac{(x - 2)^4}{32} - \frac{(x - 2)^5}{64}$; plotting f and P_5 together near $x = 2$ shows that the approximation error is less than 0.01 on $[1.06, 3.12]$ (or on any smaller interval centered at $x = 2$).

19. Finding values and derivatives at $x = 1$ gives
$P_5(x) = (x - 1) - \frac{(x - 1)^2}{2} + \frac{(x - 1)^3}{3} - \frac{(x - 1)^4}{4} + \frac{(x - 1)^5}{5}$; plotting f and P_5 together near $x = 1$ shows that the approximation error is less than 0.01 on $[0.45, 1.67]$ (or on any smaller interval centered at $x = 1$).

20. Finding values and derivatives at $x = 1$ gives
$P_5(x) = 1 - 2(x - 1) + 3(x - 1)^2 - 4(x - 1)^3 + 5(x - 1)^4 + 6(x - 1)^5$; plotting f and P_5 together near $x = 1$ shows that the approximation error is less than 0.01 on $[0.69, 1.36]$ (or on any smaller interval centered at $x = 1$).

21. Finding values and derivatives at $x = 1$ gives
$P_5(x) = 1 + \frac{x - 1}{2} - \frac{(x - 1)^2}{8} + \frac{(x - 1)^3}{16} - \frac{5(x - 1)^4}{128} + \frac{7(x - 1)^5}{256}$; plotting f and P_5 together near $x = 1$ shows that the approximation error is less than 0.01 on $[0.25, 1.97]$ (or on any smaller interval centered at $x = 1$).

22. Finding values and derivatives at $x = 4$ gives
$P_5(x) = 2 + \frac{x - 4}{4} - \frac{(x - 4)^2}{64} + \frac{(x - 4)^3}{512} - \frac{5(x - 4)^4}{16384} + \frac{7(x - 4)^5}{131072}$; plotting f and P_5 together near $x = 4$ shows that the approximation error is less than 0.01 on $[1.24, 7.44]$ (or on any smaller interval centered at $x = 4$).

23. Finding values and derivatives at $x = 1$ gives
$P_5(x) = 1 - \frac{x - 1}{2} + 3\frac{(x - 1)^2}{8} - 5\frac{(x - 1)^3}{16} + 35\frac{(x - 1)^4}{128} - 63\frac{(x - 1)^5}{256}$; plotting f and

P_5 together near $x = 1$ shows that the approximation error is less than 0.01 on [0.47, 1.64] (or on any smaller interval centered at $x = 1$).

24. Finding values and derivatives at $x = 4$ gives
$$P_5(x) = \frac{1}{2} - \frac{x-4}{16} + \frac{3(x-4)^2}{256} - \frac{5(x-4)^3}{2048} + \frac{35(x-4)^4}{65536} - \frac{63(x-4)^5}{524288}; \text{ plotting } f$$
and P_5 together near $x = 4$ shows that the approximation error is less than 0.01 on [1.66, 6.91] (or on any smaller interval centered at $x = 4$).

25. No. $P_n(x)$ involves powers of $(x - x_0)^k$ with k up to but not exceeding n.

26. Yes. $P_n(x)$ involves powers of $(x - x_0)^k$ with k up to n, but the coefficient of $(x - x_0)^n$ might be zero.

27. $P_2(x)$ is the sum of terms of $P_5(x)$ up through degree 2.

28.   (a) If f is even, then $f(-x) = f(x)$; differentiating both sides gives $-f'(-x) = f'(x)$, which means that f' is odd. Note that, for similar reasons, the derivative of an odd function is even.

     (b) If f is even, then we know f' is odd, f'' is even, f''' is odd, etc. In particular, all odd-order derivatives are odd functions. Recall that if g is any odd function, then $g(0) = g(-0) = -g(0)$, so $g(0) = 0$. Thus all odd-order derivatives are zero at zero, so every Maclaurin polynomial has only even-order terms.

29. If f is odd, then f' is even, f'' is odd, f''' is even, etc. Thus all even-order derivatives of f are odd functions, and so have the value zero for input zero. This implies that all even-order Maclaurin polynomial terms are zero.

30. Because $g(x) = f(x - b)$, we have $g(b) = f(0)$, $g'(b) = f'(0)$, $g''(b) = f''(0)$, $g'''(b) = f'''(0)$, etc. This means that all Taylor coefficients for g at $x = b$ are the same as the Maclaurin coefficients for f at 0, which is what we wanted to show. $g'(b) = f'(0)$, $g''(b) = f''(0)$, $g'''(b) = f'''(0)$, etc.

31. Note that $ln(x) = f(x - 1)$; this fact and Exercise 30 imply that the desired polynomial is $P_4(x) = (x - 1) - (x - 1)^2/2 + (x - 1)^3/3 - (x - 1)^4/4$.

32. Note that $g(x) = f(x + 1) = f(x - (-1))$; this fact and Exercise 30 imply that the desired polynomial is $P_4(x) = 1 - x + x^2 - x^3 + x^4$.

33. Note that $g(x) = f(x + 1) = f(x - (-1))$; this fact and Exercise 30 imply that the desired polynomial is $P_4(x) = 1 - x/2 + 3x^2/8 - 5x^3/16 + 35x^4/128$.

34. Since $f(x)$ and $M_n(x)$ have the same value and first n derivatives at $x = 0$, the same is true of $f(kx)$ and $M_n(kx)$. Thus $M_n(kx)$ is the nth-order Maclaurin polynomial for $f(kx)$.

35. For $f(x) = \sin x$ we have $M_5(x) = x - \dfrac{x^3}{6} + \dfrac{x^5}{120}$. Setting $k = 1/2$ in the result of Exercise 34 gives the new 5th-order Maclaurin polynomial

$$P_5(x) = \frac{x}{2} - \frac{x^3}{6 \cdot 8} + \frac{x^5}{120 \cdot 32} = \frac{x}{2} - \frac{x^3}{48} + \frac{x^5}{3840}.$$

36. For $f(x) = \cos x$ we have $M_5(x) = 1 - \dfrac{x^2}{2} + \dfrac{x^4}{24}$. Setting $k = \pi$ in the result of Exercise 34 gives the new 5th-order Maclaurin polynomial $P_5(x) = 1 - \dfrac{\pi^2 x^2}{2} + \dfrac{\pi^4 x^4}{24}$.

37. For $f(x) = e^x$ we have $M_5(x) = 1 + x + \dfrac{x^2}{2} + \dfrac{x^3}{6} + \dfrac{x^4}{24} + \dfrac{x^5}{120}$, so replacing x with $-x$ gives the new 5th-order Maclaurin polynomial $P_5(x) = 1 - x + \dfrac{x^2}{2} - \dfrac{x^3}{6} + \dfrac{x^4}{24} - \dfrac{x^5}{120}$.

38. For $f(x) = \ln(1 + x)$ we have $M_5(x) = x - \dfrac{x^2}{2} + \dfrac{x^3}{3} - \dfrac{x^4}{4} + \dfrac{x^5}{5}$. Setting $k = 2$ in the result of Exercise 34 gives the new 5th-order Maclaurin polynomial

$$P_5(x) = 2x - \frac{4x^2}{2} + \frac{8x^3}{3} - \frac{16x^4}{4} + \frac{32x^5}{5} = 2x - 2x^2 + \frac{8x^3}{3} - 4x^4 + \frac{32x^5}{5}.$$

39. For $g(x) = e^x$ we have $M_3(x) = 1 + x + \dfrac{x^2}{2} + \dfrac{x^3}{6}$; replacing x with x^2 gives

$$P_6(x) = 1 + x^2 + \frac{x^4}{2} + \frac{x^6}{6},$$ the sixth-order Maclaurin polynomial for $f(x) = e^{x^2}$.

40. For $g(x) = \sin x$ we have $M_3(x) = x - \dfrac{x^3}{6}$; replacing x with x^2 gives $P_6(x) = x^2 - \dfrac{x^6}{6}$, the sixth-order Maclaurin polynomial for $f(x) = \sin(x^2)$.

41. For $g(x) = 1/(1 + x)$ we have $M_3(x) = 1 - x + x^2 - x^3$; replacing x with x^2 gives $P_6(x) = 1 - x^2 + x^4 - x^6$, the sixth-order Maclaurin polynomial for $f(x) = 1/(1 + x^2)$.

42. For $g(x) = \sqrt{1 + x}$ we have $M_2(x) = 1 + \dfrac{x}{2} - \dfrac{x^2}{8}$. Replacing x with x^3 gives

$$P_6(x) = 1 + \frac{x^3}{2} - \frac{x^6}{8},$$ the sixth-order Maclaurin polynomial for $f(x) = \sqrt{1 + x^3}$.

43. (a) Differentiating both sides of $g(x) = xf(x)$ repeatedly (using the product rule) shows the pattern: $g'(x) = xf'(x) + f(x)$,
$g''(x) = xf''(x) + f'(x) + f'(x) = xf''(x) + 2f'(x)$, $g'''(x) = xf'''(x) + 3f''(x)$, and so on.

   (b) Part (a) implies that $xM_n(x)$ and $xf(x)$ have the same value and first $n + 1$ derivatives at $x = 0$. This means that $xM_n(x)$ is indeed the Maclaurin polynomial of order $n + 1$ for $xf(x)$.

44. The fifth-degree Maclaurin polynomial for $g(x) = \sin x$ is $M_5(x) = x - \dfrac{x^3}{6} + \dfrac{x^5}{120}$; multiplying by x gives $P_6(x) = x^2 - \dfrac{x^4}{6} + \dfrac{x^6}{120}$, the sixth-order Maclaurin polynomial for $f(x) = x \sin x$.

45. By the sum rule for derivatives, $h^{(k)}(x_0) = f^{(k)}(x_0) + g^{(k)}(x_0)$ for all orders k; similarly, $r^{(k)}(x_0) = p^{(k)}(x_0) + q^{(k)}(x_0)$. Thus, the value and first n derivatives of h agree with those of r at $x = x_0$.

46. The trigonometric identity implies that $\cos(x + \pi/3) = \dfrac{1}{2}\cos x - \dfrac{\sqrt{3}}{2}\sin x$. The fifth-order Maclaurin for the right side (and hence for $\cos(x + \pi/3)$ is therefore $\dfrac{1}{2}\left(1 - \dfrac{x^2}{2} + \dfrac{x^4}{24}\right) - \dfrac{\sqrt{3}}{2}\left(x - \dfrac{x^3}{6} + \dfrac{x^5}{120}\right)$. Now Exercise 30 says that the nth-order Taylor polynomial for $\cos x$, based at $x_0 = \pi/3$, is

$$\frac{1}{2}\left(1 - \frac{(x-\pi/3)^2}{2} + \frac{(x-\pi/3)^4}{24}\right) - \frac{\sqrt{3}}{2}\left((x-\pi/3) - \frac{(x-\pi/3)^3}{6} + \frac{(x-\pi/3)^5}{120}\right)$$

$$= \frac{1}{2} - \frac{\sqrt{3}(x-\pi/3)}{2} - \frac{(x-\pi/3)^2}{4} + \frac{\sqrt{3}(x-\pi/3)^3}{12} + \frac{(x-\pi/3)^4}{48} - \frac{\sqrt{3}(x-\pi/3)^5}{240}.$$

47. Note that $F(x)$ and $P_n(x)$ have the same value and derivatives through order n at x_0. It follows that $F'(x)$ and $P_n'(x)$ have the same value and derivatives through order $n - 1$ at x_0. Thus $P_n'(x)$ is the desired Taylor polynomial.

48. Write $Q(x) = F(x_0) + \displaystyle\int_{x_0}^{x} P_n(t)\,dt$. We need to show that (i) $Q(x_0) = F(x_0)$; and (ii) $Q^{(k)}(x_0) = F^{(k)}(x_0)$ if $1 \le k \le n + 1$. Part (i) is obvious from the definition of Q. To show (ii), note that since $F' = f$, we have $F^{(k)}(x_0) = f^{(k-1)}(x_0)$ when $1 \le k \le n+1$, and that $f^{(k-1)}(x_0) = P_n^{(k-1)}(x_0)$ for the same k. Finally, the fundamental theorem of calculus says that $Q'(x) = P_n(x)$ for all x; it follows that $Q^{(k)}(x_0) = P_n^{(k-1)}(x_0) = f^{(k-1)}(x_0) = F^{(k)}(x_0)$ holds for $1 \le k \le n + 1$, as desired.

49. Note that $\cos x$ has 6th-order Maclaurin polynomial $P_6(x) = 1 - \dfrac{x^2}{2} + \dfrac{x^4}{24} - \dfrac{x^6}{720}$. It follows from this and the hint that $\cos^2 x$ has 6th-order Maclaurin polynomial

$$Q_6(x) = \frac{1 + P_6(2x)}{2} = 1 - x^2 + \frac{x^4}{3} - \frac{2x^6}{45}.$$

50. Note that $\sin x$ has 6th-order Maclaurin polynomial $P_6(x) = x - \dfrac{x^3}{6} + \dfrac{x^5}{120}$. It follows from this and the hint that $\sin x \cos x$ has 6th-order Maclaurin polynomial

$$Q_6(x) = \frac{P_6(2x)}{2} = x - \frac{2x^3}{3} + \frac{2x^5}{15}.$$

51. Note that $\ln(1 + x)$ has 6th-order Maclaurin polynomial
    $P_6(x) = x - \dfrac{x^2}{2} + \dfrac{x^3}{3} - \dfrac{x^4}{4} + \dfrac{x^5}{5} - \dfrac{x^6}{6}$. Replacing x with $-x$ shows that $\ln(1 - x)$ has
    6th-order Maclaurin polynomial $Q_6(x) = -x - \dfrac{x^2}{2} - \dfrac{x^3}{3} - \dfrac{x^4}{4} - \dfrac{x^5}{5} - \dfrac{x^6}{6}$. Subtracting
    these polynomials gives $R_6(x) = P_6(x) - Q_6(x) = 2x + \dfrac{2x^3}{3} + \dfrac{2x^5}{5}$, the 6th-order
    Maclaurin polynomial for the original function.

52. Note that e^x has 3rd-order Maclaurin polynomial $P_3(x) = 1 + x + \dfrac{x^2}{2} + \dfrac{x^3}{6}$; therefore e^{-x^2}
    has 6th-order Maclaurin polynomial $Q_6(x) = 1 - x^2 + \dfrac{x^4}{2} - \dfrac{x^6}{6}$. Antidifferentiating Q_6
    gives $R_7(x) = x - \dfrac{x^3}{3} + \dfrac{x^5}{10} - \dfrac{x^7}{42}$, the 7th-order Maclaurin polynomial for $f(x)$. (The
    6th-order Maclaurin polynomial omits the last term.)

53. Note that $g(x) = 1/(1 + x^2)$ has 5th-order Maclaurin polynomial $P_5(x) = 1 - x^2 + x^4$.
    Integrating this gives $Q_6(x) = x - \dfrac{x^3}{3} + \dfrac{x^5}{5}$, the 6th-order Maclaurin polynomial for
    $\arctan x$.

54. Note that $g(x) = 1/\sqrt{1 - x^2}$ has 5th-order Maclaurin polynomial $P_5(x) = 1 + \dfrac{x^2}{2} + \dfrac{3x^4}{8}$.
    Integrating this gives $Q_6(x) = x + \dfrac{x^3}{6} + \dfrac{3x^5}{40}$, the 6th-order Maclaurin polynomial for
    $\arcsin x$.

§9.2 Taylor's Theorem: Accuracy Guarantees for Taylor Polynomials

1. (a) The graph shows that $|f(x) - P_1(x)|$ is largest at $x = 80$, where its value is about 0.056.

    (b) Yes; the result seen in (a) is considerably less than the bound 0.091.

2. (a) Since $f(x) = e^x$, we have $f^{(6)}(x) = e^x$ also; note $|e^x| \le e^2 \approx 7.4 < 8$ for x in the interval $[-2, 2]$.

    (b) Theorem 2, with $K_6 = 8$ and $x_0 = 0$, says that $|f(x) - P_5(x)| \le \frac{8}{6!}|x|^6 = \frac{1}{90}|x|^6$ if $-2 \le x \le 2$.

    (c) If x is in the interval $[-2, 2]$, then $|x|^6 \le 64$, so (b) implies that
    $$|f(x) - P_5(x)| \le \frac{1}{90}|x|^6 \le \frac{64}{90} \approx 0.711.$$

    (d) Plotting $|f(x) - P_5(x)|$ for $-2 \le x \le 2$ shows that the maximum approximation error occurs at $x = 2$, where $|f(x) - P_5(x)| \approx 0.122$.

    (e) The result of part (c), 0.122 is less than the upper bound, 0.72 guaranteed by the theorem, so the results are consistent.

3. (a) Note that if $f(x) = \sin x$, then $f^{(6)}(x) = -\sin x$, and so $\left|f^{(6)}(x)\right| \le 1$ for all x. Thus we can use $K_6 = 1$ in Theorem 2, which says that the approximation error $|f(x) - P_5(x)|$ doesn't exceed $\frac{1}{6!}|x|^6$ for x in $[-2, 2]$. This quantity is largest when $x = 2$; in that case $\frac{1}{6!}|x|^6 = \frac{2^6}{6!} = \frac{4}{45} \approx 0.089$.

    (b) Plotting $|f(x) - P_5(x)|$ for $-2 \le x \le 2$ shows that the maximum approximation error occurs at $x = \pm 2$, where $|f(x) - P_5(x)| \approx 0.024$.

4. (a) Note that if $f(x) = \cos x$, then $f^{(7)}(x) = -\sin x$, and so $\left|f^{(7)}(x)\right| \le 1$ for all x. Thus we can use $K_7 = 1$ in Theorem 2, which says that the approximation error $|f(x) - P_6(x)|$ doesn't exceed $\frac{1}{7!}|x|^7$ for x in $[-2, 2]$. This quantity is largest when $x = 2$; in that case $\frac{1}{7!}|x|^7 = \frac{2^7}{7!} = \frac{8}{315} \approx 0.0254$.

    (b) Plotting $|f(x) - P_6(x)|$ for $-2 \le x \le 2$ shows that the maximum approximation error occurs at $x = \pm 2$, where $|f(x) - P_6(x)| \approx 0.0061$.

5. (a) If $f(x) = \dfrac{1}{\sqrt{x}}$, then $f^{(5)}(x) = -\dfrac{945}{32\,x^{11/2}}$. For x in $[1/2, 3/2]$,
    $|f^{(5)}(x)| \le |f^{(5)}(1/2)| \approx 1336.5$. Thus we can use $K_5 = 1336.5$ in Theorem 2, which says that the approximation error $|f(x) - P_4(x)|$ doesn't exceed $\frac{1336.5}{5!}|x - 1|^5$ for x in $[1/2, 3/2]$. This quantity is largest when $|x - 1| = 1/2$; in that case
    $$\frac{1336.5}{5!}(1/2)^5 \approx 0.348.$$

    (b) Plotting $|f(x) - P_4(x)|$ for $1/2 \le x \le 3/2$ shows that the maximum approximation error occurs at $x = 1/2$, where $|f(x) - P_4(x)| \approx 0.0143$.

6. (a) If $f(x) = \ln x$, then $f^{(5)}(x) = \dfrac{24}{x^5}$. For x in $[1/2, 3/2]$, $|f^{(5)}(x)| \le |f^{(5)}(1/2)| = 768$. Thus we can use $K_5 = 768$ in Theorem 2, which says that the approximation error $|f(x) - P_4(x)|$ doesn't exceed $\frac{768}{5!}|x - 1|^5$ for x in $[1/2, 3/2]$. This quantity is largest when $|x - 1| = 1/2$; in that case $\dfrac{768}{5!}(1/2)^5 = 0.2$.

    (b) Plotting $|f(x) - P_4(x)|$ for $1/2 \le x \le 3/2$ shows that the maximum approximation error occurs at $x = 1/2$, where $|f(x) - P_4(x)| \approx 0.01$.

7. Since $f(x) = \sin x$, all derivatives of f are sines or cosines, which never exceed 1 in absolute value. Hence we can always use $K_{n+1} = 1$ in Theorem 2, which then says that $|f(x) - M_n(x)| \le \dfrac{|x|^{n+1}}{(n + 1)!}$. For any fixed x the right side goes to zero as $n \to \infty$, and so is less than 10^{-6} for large n.

8. Since $f(x) = e^x$, every derivative of f is just f again. Thus, for any x we have $\left|f^{(n+1)}(t)\right| \le e^{|x|}$ for all t in the interval $[-|x|, |x|]$ (which contains the base point, zero). Thus we can use $K_{n+1} = e^{|x|}$ in Theorem 2, which then says that $|f(x) - M_n(x)| \le \dfrac{e^{|x|}|x|^{n+1}}{(n + 1)!}$. For any fixed x, the right side goes to zero as $n \to \infty$, and so is less than 10^{-6} for large n.

9. (a) The facts that $f(0) = 0$, $f'(0) = 0$, and $f''(0) = 2$ (all are easy to check) imply that $P_2(x) = x^2$.

    (b) Here $f'''(x) = -8x^3 \cos(x^2) - 12x \sin(x^2)$; a graph shows that $|f'''(x)| < 2.5$ for x in $[0, 1/2]$. Now Theorem 2, with $K_3 = 2.5$, gives $\left|\sin(t^2) - t^2\right| \le 2.5t^3/3!$.

    (c) The result of part (b) implies that

    $$\left|\int_0^{1/2} \sin(x^2)\,dx - \frac{1}{24}\right| = \left|\int_0^{1/2} \left(\sin(x^2) - x^2\right)dx\right|$$
    $$\le \int_0^{1/2} \left|\sin(x^2) - x^2\right|\,dx \le \int_0^{1/2} 2.5x^3/6\,dx$$
    $$= \frac{5}{768} \approx 0.0065.$$

    (d) The given inequality implies that

    $$\left|\int_0^{1/2} \sin(x^2)\,dx - \int_0^{1/2} x^2\,dx\right| \le \int_0^{1/2} \left|\sin(x^2) - x^2\right|\,dx$$
    $$\le \int_0^{1/2} 120x^6/6!\,dx \approx 0.00019.$$

10. If $f(x) = e^{-x^2}$, then $P_2(x) = 1 - x^2$. Also, $f'''(x) = \dfrac{12x}{e^{x^2}} - \dfrac{8x^3}{e^{x^2}}$; and a graph shows that $|f'''(x)| \le 4$ on the interval $[0, 2/3]$, Now Taylor's theorem guarantees that

$|f(x) - P_2(x)| \le \dfrac{4}{3!}x^3$ on the interval $[0, 2/3]$, Therefore,

$$\left| \int_0^{2/3} f(x)\, dx - \int_0^{2/3} P_2(x)\, dx \right| \le \int_0^{2/3} |f(x) - P_2(x)|\, dx \le \int_0^{2/3} \frac{4}{3!}x^3\, dx \approx 0.033.$$

11. Theorem 2 implies that $|x - \sin x| \le \dfrac{K_3}{3!}|x|^3$, where K_3 is a constant. (In fact, we can use $K_3 = 1$, but all that matters here is that K_3 is a constant.) Thus,

$$\left| \frac{x - \sin x}{x^2} \right| \le \frac{K_3 |x|^3}{3!\, x^2} = \frac{K_3}{3!}|x|.$$

It follows that $\lim\limits_{x \to 0} \left| \dfrac{x - \sin x}{x^2} \right| = 0$ and so $\lim\limits_{x \to 0} \dfrac{x - \sin x}{x^2} = 0.$

12. Theorem 2 implies that $|x - \sin x| \le \dfrac{K_2}{2}|x|^2$, where K_2 is a constant. Thus,

$$\left| \frac{x - \sin x}{x} \right| \le \frac{K_2 |x|^2}{2\,|x|} = \frac{K_2}{2}|x|.$$

Thus, $\lim\limits_{x \to 0} \dfrac{x - \sin x}{x} = 0.$

13. Theorem 2 implies that $\left| \cos x - 1 + x^2/2 \right| \le \dfrac{K_4}{4!}|x|^4$, where K_4 is a constant. Thus,

$$\left| \frac{\cos x - 1 + x^2/2}{x^3} \right| \le \frac{K_4 |x|^4}{4!\, |x|^3} = \frac{K_4}{4!}|x|.$$

Thus, $\lim\limits_{x \to 0} \dfrac{\cos x - 1 + x^2/2}{x^3} = 0.$

14. Theorem 2 implies that $\left| e^x - 1 - x \right| \le \dfrac{K_2}{2}|x|^2$, where K_2 is a constant. Thus,

$$\left| \frac{e^x - 1 - x}{x} \right| \le \frac{K_2 |x|^2}{2\,|x|} = \frac{K_2}{2}|x|.$$

Thus, $\lim\limits_{x \to 0} \dfrac{e^x - 1 - x}{x} = 0.$

15. (a) $\sqrt{1 + x} \approx 1 + \dfrac{1}{2}x - \dfrac{1}{8}x^2.$

(b) $V = 2\pi\sigma\left(\sqrt{r^2 + a^2} - r \right) = 2\pi\sigma r \left(\sqrt{1 + \left(\dfrac{a}{r}\right)^2} - 1 \right)$

$$\approx 2\pi\sigma r \left(\frac{a^2}{2r^2} - \frac{a^4}{8r^4} \right) = 2\pi\sigma \left(\frac{a^2}{2r} - \frac{a^4}{8r^3} \right)$$

16. Consider $m = \dfrac{m_0}{\sqrt{1 - (v/c)^2}}$ as a function of v. Differentiation gives $m(0) = m_0$, $m'(0) = 0$,

    and $m''(0) = m_0/c^2$; thus the second-order Maclaurin polynomial for m is

    $$P_2(x) = m_0 + m_0 \frac{v^2}{2c^2}.$$

17. (a) $f(1.5) \approx f(1) + f'(1) \cdot (1.5 - 1) + \dfrac{f''(1)}{2}(1.5 - 1)^2$

    $$= 1 + 2 \cdot (0.5) + \frac{1}{4}(0.5)^2 = \frac{33}{16} = 2.0625.$$

    (b) Since $f'''(x) = -\dfrac{3x^2}{(1 + x^3)^2}$, $K_3 = 3/4$. Therefore, the approximation error does not

    exceed $\dfrac{K_3}{3!}(1.5 - 1)^3 = \dfrac{1}{64} = 0.015625.$

18. From the information given,

    $$f(x) = 26 + 22x - \frac{16}{2!}x^2 + \frac{12}{3!}x^3 + \cdots = 26 + 22x - 8x^2 + 2x^3 + \ldots. \text{ Since}$$

    $P_3(1) = 42$, Theorem 2 implies that $|f(x) - 42| < \dfrac{7 \cdot 81}{4!} = \dfrac{189}{8}.$

    (a) $f(1) \leq 42 + \dfrac{189}{8} = \dfrac{525}{8}.$

    (b) $f(1) \geq 42 - \dfrac{189}{8} = \dfrac{147}{8}.$

19. (a) By the fundamental theorem of calculus, $\displaystyle\int_a^x f'(t)\,dt = f(x) - f(a)$, and the result
    follows.

    (b) Let $u = f'(t)$, $dv = dt$, $du = f''(t)\,dt$ and $v = t - x$. (Letting $v = t$ is also possible,
    but $v = t - x$ works better here.) Then

    $$f(x) = f(a) + f'(t)(t - x)\Big|_a^x - \int_a^x (t - x) f''(t)\,dt$$

    $$= f(a) - f'(a)(a - x) + \int_a^x (x - t) f''(t)\,dt.$$

    (c) Let $u = f''(t)$ and $dv = (x - t)\,dt$. Then, $du = f'''(t)\,dt$ and $v = -(x - t)^2/2$. Thus,

    $$f(x) = f(a) + f'(a)(x - a) - \frac{1}{2}f''(t)(x - t)^2\Big|_a^x + \frac{1}{2}\int_a^x (x - t)^2 f'''(t)\,dt$$

    $$= f(a) + f'(a)(x - a) + \frac{1}{2}f''(a)(x - a)^2 + \frac{1}{2}\int_a^x (x - t)^2 f'''(t)\,dt.$$

    This is the desired formula for $n = 2$.

Now, applying integration by parts with $u = f'''(t)$, $dv = (x - t)^2/2\, dt$, $du = f^{(4)}(t)\, dt$, $v = -(x - t)^3/3!$ gives (by a similar calculation)

$$f(x) = f(a) + f'(a)(x-a) + \frac{1}{2}f''(a)(x-a)^2 + \frac{1}{3!}f'''(a)(x-a)^3 + \frac{1}{3!}\int_a^x (x-t)^3 f^{(4)}(t)\, dt.$$

The same pattern persists for larger n.

20. Assume for simplicity that $x > 0$ (the case $x < 0$ is almost the same). Because $\left|f^{(n+1)}(t)\right| \le K_{n+1}$ for all t, we have

$$|f(x) - P_n(x)| = |R_n(x)| = \left| \frac{1}{n!} \int_0^x (x - t)^n f^{(n+1)}(t)\, dt \right|$$

$$\le \frac{K_{n+1}}{n!} \int_0^x (x - t)^n \, dt = -\frac{K_{n+1}}{n!} \left. \frac{(x - t)^{n+1}}{n + 1} \right|_0^x = \frac{K_{n+1}}{(n + 1)!} x^{n+1}.$$

§9.3   Fourier Polynomials: Approximating Periodic Functions

1. (a) Both $\sin(x)$ and $\cos(x)$ are 2π-periodic. Thus
    $f(x + 2\pi) = \cos(k(x + 2\pi)) = \cos(kx + 2k\pi) = \cos(kx) = f(x)$. The function g
    behaves the same way.

    (b) Each summand of a trigonometric polynomial is 2π-periodic. Therefore, so is the
    polynomial itself.

    (c) The smallest period of $p(x) = \cos(4x) + \sin(8x)$ is $\pi/2$, because
    $p(x + \pi/2) = \cos(4(x + \pi/2)) + \sin(8(x + \pi/2)) = \cos(4x + 2\pi) + \sin(8x + 4\pi) =$
    $\cos(4x) + \sin(8x) = p(x)$.

2. (a) $\displaystyle\int_{-\pi}^{\pi} \cos(kx)\, dx = \left.\frac{\sin(kx)}{k}\right|_{-\pi}^{\pi} = 0$ since $\sin(k\pi) = 0$ for any integer k.

    (b) The function $\sin(kx)$ is odd, so $\displaystyle\int_{-a}^{a} \sin(kx)\, dx = 0$ for any real number a.

3. (a) Let $u = \sin(mx)$. Then, $\displaystyle\int_{-\pi}^{\pi} \cos(mx)\sin(mx)\, dx = \frac{1}{m}\int_{0}^{0} u\, du = 0$.

    (b) The integrand is odd because it's the product of an odd function and an even function.
    Integrating any odd function over $[-\pi, \pi]$ yields zero.

    (c) $\displaystyle\int_{-\pi}^{\pi} \cos(mx)\cos(nx)\, dx = \frac{1}{2}\int_{-\pi}^{\pi} \cos((m + n)x)\, dx$
    $$+ \frac{1}{2}\int_{-\pi}^{\pi} \cos((m - n)x)\, dx = 0 + 0$$
    by the result in part (a) of the previous exercise.

    (d) $\displaystyle\int_{-\pi}^{\pi} \cos^2(mx)\, dx = \left.\left(\frac{x}{2} + \frac{\sin(2mx)}{4m}\right)\right|_{-\pi}^{\pi} = \pi$.

    (e) $\displaystyle\int_{-\pi}^{\pi} \sin(mx)\sin(nx)\, dx = \frac{1}{2}\int_{-\pi}^{\pi} \cos((m - n)x)\, dx - \frac{1}{2}\int_{-\pi}^{\pi} \cos((m + n)x)\, dx$
    $$= 0 - 0$$
    by the result in part (a) of the previous exercise.

    (f) $\displaystyle\int_{-\pi}^{\pi} \sin^2(mx)\, dx = \left.\left(\frac{x}{2} - \frac{\sin(2mx)}{4m}\right)\right|_{-\pi}^{\pi} = \pi$.

4. All summands give zero integrals except the constant term.

5. All summands give zero integrals except the $a_k \cos(kx)$ term.

6. (a) Because f is even, $f(-x) = f(x)$. Because g is odd, $g(-x) = -g(x)$. Thus,
    $h(-x) = f(-x) \cdot g(-x) = f(x) \cdot -g(x) = -f(x) \cdot g(x) = -h(x)$. Thus h is odd.

    (b) Since both f and g are even, we have $h(-x) = f(-x) \cdot g(-x) = f(x) \cdot g(x) = h(x)$,
    so h is even.

(c) Since both f and g are odd, we have

$$h(-x) = f(-x) \cdot g(-x) = -f(x) \cdot -g(x) = f(x) \cdot g(x) = h(x), \text{ so } h \text{ is even.}$$

7. For all $k > 0$ we have $b_k = \dfrac{1}{\pi} \displaystyle\int_{-\pi}^{\pi} f(x) \sin(kx)\, dx$. Here the integrand is an odd function, since it's the product of an even function, $f(x)$, and an odd function, $\sin(kx)$. Integrating an odd function over $[-\pi, \pi]$ gives zero.

8. For $k > 0$ we have $a_k = \dfrac{1}{\pi} \displaystyle\int_{-\pi}^{\pi} f(x) \cos(kx)\, dx$. Here the integrand is an odd function, since it's the product of an odd function, $f(x)$, and an even function, $\cos(kx)$. Integrating an odd function over $[-\pi, \pi]$ gives zero. For $k = 0$, we have $a_0 = \dfrac{1}{2\pi} \displaystyle\int_{-\pi}^{\pi} f(x)\, dx$. Since the integrand is odd, the integral is zero.

9. (a) The Fourier polynomial of degree 2 is $q_2(x) = 2/\pi$—the sum of terms through degree two in $q_5(x)$.

   (b) The Fourier polynomial of degree 3 is $q_3(x) = 2/\pi + 4\sin(3x)$—the sum of terms through degree three in $q_5(x)$.

   (c) $q_4(x) = 2/\pi + 4\sin(3x)$—the sum of terms through degree in $q_5(x)$.

10. For any n, the Fourier polynomial of degree n for f is just the sum of terms through degree n in f.

   (a) Terms through $n = 3$ in f give $q_3(x) = 0$, since there are no terms of degree ≤ 3 in f.

   (b) $q_4(x) = 3\sin(4x)$

   (c) Since f has no terms of degree greater than 5, neither can q_n.

11. The average value of f over $[-\pi, \pi]$ is $\dfrac{1}{2\pi} \displaystyle\int_{-\pi}^{\pi} f(x)\, dx$—exactly the same as the constant term a_0. Here the constant term is 3.

12. The Fourier polynomial for $f(x) + 1$ is the sum of the Fourier polynomial for $f(x)$ and the Fourier polynomial for 1. The latter is just 1. Therefore, the Fourier polynomial of degree n for g is simply $1 + q_n$.

13. The key point is that each Fourier coefficient for the sum function $f + g$ is the sum of the corresponding Fourier coefficients for the separate functions f and g. To see why this works for the sine coefficients a_k, we'll write $a_{k,f}$, $a_{k,g}$, $a_{k,f+g}$ for the kth sine coefficient of f, g, and $f + g$, respectively. Then

$$a_{k,f+g} = \frac{1}{\pi} \int_{-\pi}^{\pi} (f(x) + g(x)) \sin(kx)\, dx$$

$$= \frac{1}{\pi} \int_{-\pi}^{\pi} f(x) \sin(kx)\, dx + \frac{1}{\pi} \int_{-\pi}^{\pi} g(x) \sin(kx)\, dx = a_{k,f} + a_{k,g}.$$

The situation is the same for the cosine coefficients: $b_{k,f+g} = b_{k,f} + b_{k,g}$.

14. $\displaystyle\int_{-\pi}^{\pi} x\,\sin(kx)\,dx = \left(\frac{1}{k^2}\sin(kx) - \frac{x}{k}\cos(kx)\right)\Big]_{-\pi}^{\pi} = -\frac{2\pi}{k}\cos(k\pi)$ so

$\displaystyle\frac{1}{\pi}\int_{-\pi}^{\pi} x\,\sin(kx)\,dx = \frac{-2\cos(k\pi)}{k}.$

15. (a) $\displaystyle a_0 = \frac{1}{2\pi}\int_{-\pi}^{\pi} f(x)\,dx = \frac{1}{2\pi}\int_{0}^{\pi} dx = \frac{1}{2}$

$\displaystyle a_k = \frac{1}{\pi}\int_{-\pi}^{\pi} f(x)\cos(kx)\,dx = \frac{1}{\pi}\int_{0}^{\pi}\cos(kx)\,dx = \frac{\sin(kx)}{k\pi}\Big]_{0}^{\pi} = 0$

$\displaystyle b_k = \frac{1}{\pi}\int_{-\pi}^{\pi} f(x)\sin(kx)\,dx = \frac{1}{\pi}\int_{0}^{\pi}\sin(kx)\,dx = -\frac{\cos(kx)}{k\pi}\Big]_{0}^{\pi} = \frac{1-\cos(k\pi)}{k\pi}.$

Thus, $b_{2m} = 0$ and $b_{2m+1} = \dfrac{2}{(2m+1)\pi}$ for $m = 0, 1, 2, 3, \ldots.$

(b) $\displaystyle q_1(x) = \frac{1}{2} + \frac{2\sin x}{\pi};$

$\displaystyle q_3(x) = \frac{1}{2} + \frac{2\sin x}{\pi} + \frac{2\sin(3x)}{3\pi};$

$\displaystyle q_5(x) = \frac{1}{2} + \frac{2\sin x}{\pi} + \frac{2\sin(3x)}{3\pi} + \frac{2\sin(5x)}{5\pi};$

$\displaystyle q_7(x) = \frac{1}{2} + \frac{2\sin x}{\pi} + \frac{2\sin(3x)}{3\pi} + \frac{2\sin(5x)}{5\pi} + \frac{2\sin(7x)}{7\pi}.$

(d) Let s be the square wave function in Example 2. Then,
$f(x) = \big(s(x) + 1\big)/2 = s(x)/2 + 1/2.$ The same relationship holds between the Fourier polynomial found in Example 2 and the one in this exercise.

16. (a) $a_0 = \dfrac{\pi}{2}$ and $a_k = \dfrac{2}{\pi k^2}\big(1 - \cos(k\pi)\big)$. Thus, $a_{2m} = 0$ and $a_{2m-1} = 4/(\pi(2m-1)^2)$ for $m = 1, 2, 3, \ldots.$

$b_k = 0$

(b) $\displaystyle q_1(x) = \frac{\pi}{2} + \frac{4\cos x}{\pi};$

$\displaystyle q_3(x) = \frac{\pi}{2} + \frac{4\cos x}{\pi} + \frac{4\cos(3x)}{9\pi};$

$\displaystyle q_5(x) = \frac{\pi}{2} + \frac{4\cos x}{\pi} + \frac{4\cos(3x)}{9\pi} + \frac{4\cos(5x)}{25\pi};$

$\displaystyle q_7(x) = \frac{\pi}{2} + \frac{4\cos x}{\pi} + \frac{4\cos(3x)}{9\pi} + \frac{4\cos(5x)}{25\pi} + \frac{4\cos(7x)}{49\pi}.$

17. Because f is an even function, all the b_k are zero. To find the a_k, note that all integrands are even functions, so (for simplicity) we can integrate from 0 to π and double the result. Note, finally, that $f(x) = |x| = x$ if $x \geq 0$. Thus, we have

$\displaystyle a_0 = \frac{1}{2\pi}\int_{-\pi}^{\pi} |x|\,dx = \frac{1}{\pi}\int_{0}^{\pi} x\,dx = \frac{\pi}{2}.$ Similarly, if $k > 0$ we have

$a_k = \dfrac{2}{\pi} \displaystyle\int_0^\pi x \cos(kx)\, dx$. Evaluating these integrals for $k = 1, 2, \ldots, 7$ gives the pattern

$$\frac{-4}{\pi}, \quad 0, \quad \frac{-4}{9\pi}, \quad 0, \quad \frac{-4}{25\pi}, \quad 0, \quad \frac{-4}{49\pi}.$$

Therefore, $q_7(x) = \dfrac{\pi}{2} - \dfrac{4}{\pi}\cos(x) - \dfrac{4}{9\pi}\cos(3x) - \dfrac{4}{25\pi}\cos(5x) - \dfrac{4}{49\pi}\cos(7x)$.

18. Because f is even, $b_k = 0$ for all k. To find the a_k, note that all integrands are even functions, so we can always integrate from 0 to π and double our results.
$a_0 = \dfrac{1}{\pi}\displaystyle\int_0^\pi x^2\, dx = \dfrac{\pi^2}{3}$. Similarly, if $k > 0$ we have $a_k = \dfrac{2}{\pi}\displaystyle\int_0^\pi x^2 \cos(kx)\, dx$.
Evaluating these integrals for $k = 1, 2, \ldots, 7$ (technology helps) gives the pattern

$$-4, \quad 1, \quad -\frac{4}{9}, \quad \frac{1}{4}, \quad -\frac{4}{25}, \quad \frac{1}{9}, \quad -\frac{4}{49}.$$

Therefore,

$$q_7(x) = \frac{\pi^2}{3} - 4\cos(x) + \cos(2x) - \frac{4\cos(3x)}{9}$$
$$+ \frac{\cos(4x)}{4} - \frac{4\cos(5x)}{25} + \frac{\cos(6x)}{9} - \frac{4\cos(7x)}{49}.$$

19. The constant function 4 is its own Fourier polynomial, and we've already found Fourier polynomials for x (which we can multiply by 3) and x^2 in earlier work. Combining these results gives

$$4 + \frac{\pi^2}{3} - 4\cos x + \cos(2x) - \frac{4}{9}\cos(3x) + \frac{1}{4}\cos(4x) - \frac{4}{25}\cos(5x) + \frac{1}{9}\cos(6x) - \frac{4}{49}\cos(7x)$$
$$+ 6\sin(x) - 3\sin(2x) + 2\sin(3x) - \frac{3}{2}\sin(4x) + \frac{6}{5}\sin(5x) - \sin(6x) + \frac{6}{7}\sin(7x).$$

20. (a) If $-\pi \le x \le \pi$, then $-\dfrac{T\pi}{2\pi} \le \dfrac{Tx}{2\pi} \le \dfrac{T\pi}{2\pi}$, or $-\dfrac{T}{2} \le \dfrac{Tx}{2\pi} \le \dfrac{T}{2}$, so g is indeed defined for such x. Also,
$$g(x + 2\pi) = f\left(\frac{T(x + 2\pi)}{2\pi}\right) = f\left(\frac{Tx}{2\pi} + T\right) = f\left(\frac{Tx}{2\pi}\right) = g(x); \text{ this shows that}$$
g is periodic.

   (b) Since $g(x) \approx q_n(x)$, replacing x with $Tx/(2\pi)$ gives the desired approximation.

   (c) Replacing $f(x)$ in each integral expression with the approximation from (b) gives the desired result.

10   *Improper Integrals*

§10.1   Improper Integrals: Ideas and Definitions

1. The interval of integration is infinite.

2. The integrand is unbounded near $x = 1$: $\displaystyle\lim_{x \to 1^-} \frac{1}{x^2 - 3x + 2} = \infty$.

3. The integrand is unbounded near $x = 1$: $\displaystyle\lim_{x \to 1^+} 1/(x^2 \ln x) = \infty$.

4. The integrand is unbounded near $x = \pi$: $\displaystyle\lim_{x \to \pi} \frac{\cos x}{\sqrt{1 + \cos x}} = -\infty$.

5. The integrand is improper at the left endpoint, so $\displaystyle\int_0^1 \frac{dx}{\sqrt{x}} = \lim_{t \to 0^+} \int_t^1 \frac{dx}{\sqrt{x}}$ if the limit exists.

   But $\displaystyle\int_t^1 \frac{dx}{\sqrt{x}} = 2\sqrt{x}\Big]_t^1 = 2 - 2\sqrt{t}$. Thus, $\displaystyle\int_0^1 \frac{dx}{\sqrt{x}} = \lim_{t \to 0^+}(2 - 2\sqrt{t}) = 2$.

6. The integrand is improper due to the infinite interval of integration. An easy calculation

   gives $\displaystyle\int_1^t \frac{du}{u^{3/2}} = 2 - \frac{2}{\sqrt{t}}$, so $\displaystyle\int_1^\infty \frac{du}{u^{3/2}} = \lim_{t \to \infty}\left(2 - \frac{2}{\sqrt{t}}\right) = 2$.

7. $\displaystyle\int_1^\infty \frac{dx}{x^3} = \lim_{t \to \infty} \int_1^t \frac{dx}{x^3} = \lim_{t \to \infty} \frac{1}{2}\left(1 - \frac{1}{t^2}\right) = \frac{1}{2}$.

8. $\displaystyle\int_0^4 \frac{dx}{\sqrt{x}} = \lim_{t \to 0^+} \int_t^4 \frac{dx}{\sqrt{x}} = \lim_{t \to 0^+} 2\sqrt{x}\Big]_t^4 = \lim_{t \to 0^+}\left(4 - 2\sqrt{t}\right) = 4$.

9. Integration by parts (with $u = \ln x$, $dv = dx/x^2$) gives $\displaystyle\int_1^t \frac{\ln x}{x^2}\,dx = 1 - \frac{1}{t} - \frac{\ln t}{t}$, so

   $\displaystyle\int_1^\infty \frac{\ln x}{x^2}\,dx = \lim_{t \to \infty}\left(1 - \frac{1}{t} - \frac{\ln t}{t}\right) = 1$. (The last part of the limit can be found by l'Hôpital's rule.)

10. $\displaystyle\int_e^\infty \frac{dx}{x(\ln x)^2} = \lim_{t \to \infty} \frac{-1}{\ln x}\Big]_e^t = \lim_{t \to \infty}\left(1 - \frac{1}{\ln t}\right) = 1$.

11. $\displaystyle\int_0^\infty e^{-x}\,dx = \lim_{t \to \infty} \int_0^t e^{-x}\,dx = \lim_{t \to \infty} -e^{-x}\Big]_0^t = \lim_{t \to \infty}\left(1 - e^{-t}\right) = 1$.

12. $\displaystyle I = \int_0^\infty xe^{-x}\,dx = \lim_{t \to \infty} \int_0^t xe^{-x}\,dx = \lim_{t \to \infty}\left(-xe^{-x} - e^{-x}\right)\Big]_0^t = \lim_{t \to \infty}\left(1 - te^{-t} - e^{-t}\right)$

    $\displaystyle = 1 - \lim_{t \to \infty}\left(te^{-t}\right) = 1$.

    (The last limit is zero by l'Hôpital's rule.)

13. $\displaystyle\int_{-\infty}^{1} e^x\,dx = \lim_{t\to-\infty}\left(e^1 - e^t\right) = e.$

14. Integration by parts gives $\displaystyle\int_{t}^{0} xe^x\,dx = e^t - te^t - 1.$ As $t \to -\infty$, the right side tends to the desired value, -1. (To see that $\lim_{t\to-\infty} te^t = 0$, one can apply l'Hôpital's rule:
$$\lim_{t\to-\infty} te^t = \lim_{t\to-\infty}\frac{t}{e^{-t}} = \lim_{t\to-\infty}\frac{1}{-e^{-t}} = 0.)$$

15. The integrand is undefined at $x = 0$, so we want $\displaystyle\lim_{t\to 0^+}\int_{t}^{16}\frac{dx}{\sqrt[4]{x^3}}$. Note that $\dfrac{1}{\sqrt[4]{x^3}} = \dfrac{1}{x^{3/4}}$;
    now a routine calculation gives $\displaystyle\int_{t}^{16}\frac{dx}{x^{3/4}} = 8 - 4\sqrt[4]{t}$; taking the limit as $t \to 0^+$ gives 8, the desired integral.

16. The integrand is undefined at $x = 1$, so we want $\displaystyle\lim_{t\to 1^-}\int_{0}^{t}\frac{x}{\sqrt{1-x^2}}\,dx$. The u-substitution
    $u = 1 - x^2$ leads to $\displaystyle\int_{0}^{t}\frac{x}{\sqrt{1-x^2}}\,dx = 1 - \sqrt{1-t^2}$; taking the limit as $t \to 1^-$ gives 1, the desired integral.

17. We want $\displaystyle\lim_{t\to\infty}\int_{3}^{t}\frac{x}{\left(x^2-4\right)^3}\,dx$. The u-substitution $u = x^2 - 4$ leads to
    $\displaystyle\int_{3}^{t}\frac{x}{\left(x^2-4\right)^3}\,dx = -\frac{1}{4(x^2-4)^2}\Bigg]_{3}^{t} = \frac{1}{100} - \frac{1}{4(t^2-4)^2}.$ Taking the limit as $t \to \infty$ gives $1/100$, the desired integral.

18. Substituting $u = \arctan x,\ du = dx/(1+x^2)$ gives
    $$\int_{0}^{\infty}\frac{\arctan x}{1+x^2}\,dx = \lim_{t\to\infty}\frac{1}{2}\left(\arctan t\right)^2 = \pi^2/8.$$

19. Yes, because $\displaystyle\int_{0}^{\infty} f(x)\,dx - \int_{0}^{100} f(x)\,dx = \int_{100}^{\infty} f(x)\,dx$, the right side is the difference of two convergent integrals, and so must converge itself.

20. Note that $\displaystyle\int_{5}^{\infty} f(x)\,dx = \int_{2}^{5} f(x)\,dx + \int_{5}^{\infty} f(x)\,dx$. Thus the left side is the sum of two convergent integrals, and so must converge itself.

21. The area between the x-axis and the curve $y = g(x)$ from $x = 1$ to $x = t$ is no more than the area between the x-axis and the curve $y = x^{-2}$ from $x = 1$ to $x = t$. The inequality implies that the first area is no greater than the second. The second area is found by integration:
    $\displaystyle\int_{1}^{t} x^{-2}\,dx = 1 - t^{-1}$, and this tends to 1 as $t \to \infty$. Hence the smaller area $\displaystyle\int_{1}^{t} g(x)\,dx$ must also tend to some finite limit (no bigger than 1) as $t \to \infty$.

22. The area between the x-axis and the curve $y = h(x)$ from $x = 1$ to $x = t$ is no less than the area between the x-axis and the curve $y = 1/x$ from $x = 1$ to $x = t$. An easy integral shows that the second area is $\ln t$. Because $\ln t \to \infty$ as $t \to \infty$, the larger area must also tend to infinity. In other words, $\displaystyle\int_1^\infty h(x)\,dx$ diverges.

23. Because f is odd we have $f(-x) = -f(x)$; it follows that $\displaystyle\int_{-\infty}^0 f(x)\,dx = -17$. Thus

$$\int_{-\infty}^\infty f(x)\,dx = \int_{-\infty}^0 f(x)\,dx + \int_0^\infty f(x)\,dx = 17 - 17 = 0.$$

24. (a) We've seen that for *any* odd function f, $\displaystyle\int_{-a}^a f(x)\,dx = 0$. (Draw a picture to see why.) Here we have $f(x) = x$, an odd function, so $\displaystyle\lim_{a\to\infty}\int_{-a}^a x\,dx = \lim_{a\to\infty} 0 = 0$.

    (b) For $\displaystyle\int_{-\infty}^\infty x\,dx$ to converge requires that both integrals $\displaystyle\int_{-\infty}^0 x\,dx$ and $\displaystyle\int_0^\infty x\,dx$ converge. In fact, both integrals *diverge*, by easy calculations.

25. (a) I is improper because the integrand is undefined at $x = 0$, (and blows up in magnitude nearby).

    (b) No; I diverges. For $\displaystyle\int_{-1}^1 \frac{1}{x^3}\,dx$ to converge requires that both improper integrals $\displaystyle\int_{-1}^0 \frac{1}{x^3}\,dx$ and $\displaystyle\int_0^1 \frac{1}{x^3}\,dx$ converge. In fact, both *diverge*, as easy calculations show.

26. The integral diverges. Note the impropriety at $x = 2$. If the integral converged, so would both $\displaystyle\int_0^2 \frac{dx}{x-2}$ and $\displaystyle\int_2^5 \frac{dx}{x-2}$ and both of these are easily seen to diverge.

27. (a) Substituting $u = \sqrt{x}$ and $du = \dfrac{1}{2\sqrt{x}}\,dx$ shows that

    $$\int_1^t \frac{\cos(\sqrt{x})}{\sqrt{x}}\,dx = 2\sin(\sqrt{t}) - 2\sin 1.$$ Now $\sin(\sqrt{t})$ has no limit as $t \to \infty$ (see Example 6) so the original integral diverges.

    (b) The integral $\displaystyle\int_0^1 \frac{\cos(\sqrt{x})}{\sqrt{x}}\,dx$ converges. If $t > 0$, then (as above)

    $$\int_t^1 \frac{\cos(\sqrt{x})}{\sqrt{x}}\,dx = 2\sin 1 - 2\sin(\sqrt{t}),$$ which tends to $2\sin 1$ as $t \to 0^+$.

28. Region R has area $\ln 2$. The area is measured by $\displaystyle\int_1^\infty \left(\frac{1}{x} - \frac{1}{x+1}\right) dx$, which converges. To see why, note that (by easy calculations)

$\int_1^t \left(\frac{1}{x} - \frac{1}{x+1}\right) dx = \ln(t) - \ln(1-t) + \ln(2)$. To find the limit as $t \to \infty$, note that

$\ln(t) - \ln(1+t) = \ln\left(\frac{t}{1+t}\right)$, and this "clearly" tends to zero as $t \to \infty$.

29. The region S has infinite area — the improper integral $\int_0^1 \left(\frac{1}{x} - \frac{1}{\sqrt{x}}\right) dx$ diverges. To see

why, note that (by straightforward calculations) $\int_t^1 \left(\frac{1}{x} - \frac{1}{\sqrt{x}}\right) dx = 2\sqrt{t} - \ln t - 2$,

which tends to infinity as $t \to 0$.

30. The arclength formula says that the graph of $y = f(x)$ from $x = a$ to $x = b$ has length

$\int_a^b \sqrt{1 + f'(x)^2}\, dx$. We'll find the length from $x = 0$ to $x = R$ and quadruple the result.

Here $f(x) = \sqrt{R^2 - x^2}$, so $f'(x) = \frac{-x}{\sqrt{R^2 - x^2}}$. Now some algebra gives

$\sqrt{1 + f'(x)^2} = \frac{R}{\sqrt{R^2 - x^2}}$; thus, the integral we want is $\int_0^R R \frac{dx}{\sqrt{R^2 - x^2}}$, which is

improper at $x = R$. But calculation shows that the integral converges to $R\pi/2$, which shows that the total circumference is $2\pi R$.

31. The solid's cross-sectional area (perpendicular to the x-axis) at any given x is $\pi r^2 = \frac{\pi}{x^2}$.

The volume is therefore given by the improper integral $\int_1^\infty \frac{\pi}{x^2}$; the value is π, as we

calculated in Example 1.

32. For any y with $0 \le y \le 1$, the solid's cross-section (perpendicular to the y-axis) at height y

is a "washer" with inner radius $r = 1$ and outer radius $R = 1/y$. This washer has area

$\pi(R^2 - r^2) = \pi(1/y^2 - 1)$, so the volume sought is given by $\pi \int_0^1 \left(\frac{1}{y^2} - 1\right) dy$. Since

this integral diverges (see Example 4) the volume in question is infinite.

33. Diverges. $\int_0^\infty \frac{x}{\sqrt{1+x^2}} dx = \lim_{t\to\infty} \left(\sqrt{1+t^2} - 1\right) = \infty$.

34. Converges. $\int_1^\infty \frac{dx}{x(1+x)} = \lim_{t\to\infty} \ln\left(\frac{x}{1+x}\right)\Big]_1^t = \lim_{t\to\infty} \ln\left(\frac{t}{1+t}\right) - \ln\frac{1}{2} = \ln 2$.

35. Converges.

$\int_{-2}^2 \frac{2x+1}{\sqrt[3]{x^2+x-6}} dx = \lim_{t\to 2^-} \int_{-2}^t \frac{2x+1}{\sqrt[3]{x^2+x-6}} dx = \lim_{t\to 2^-} \frac{3}{2}\left(x^2+x-6\right)^{2/3}\Big]_{-2}^t$

$= \lim_{t\to 2^-} \frac{3}{2}\left((t^2+t-6)^{2/3} - (-4)^{2/3}\right) = -\frac{3\sqrt[3]{16}}{2} = -3\sqrt[3]{2}$.

36. Converges. $\int_\pi^\infty e^{-x}\sin x\, dx = \lim_{t\to\infty} -\frac{1}{2}e^{-x}(\sin x + \cos x)\Big]_\pi^t = -\frac{1}{2}e^{-\pi}$.

37. Converges. $\displaystyle\int_2^4 \frac{x}{\sqrt{|x^2-9|}}\,dx = \int_2^3 \frac{x}{\sqrt{9-x^2}}\,dx + \int_3^4 \frac{x}{\sqrt{x^2-9}}\,dx$

$$= \lim_{s\to 3^-} -\sqrt{9-x^2}\Big]_2^s + \lim_{t\to 3^+}\sqrt{x^2-9}\Big]_t^4 = \sqrt{5}+\sqrt{7}.$$

38. Converges. $\displaystyle\int_1^3 \frac{dx}{\sqrt[3]{x-2}} = \lim_{s\to 2^-}\int_1^s \frac{dx}{\sqrt[3]{x-2}} + \lim_{t\to 2^+}\int_t^3 \frac{dx}{\sqrt[3]{x-2}}$

$$= \lim_{s\to 2^-}\frac{3}{2}(x-2)^{2/3}\Big|_1^s + \lim_{t\to 2^+}\frac{3}{2}(x-2)^{2/3}\Big|_t^3 = -\frac{3}{2}+\frac{3}{2}=0.$$

39. Converges. $\displaystyle\int_2^3 \frac{x}{\sqrt{3-x}}\,dx = \lim_{t\to 3^-}\left(\frac{2}{3}(3-t)^{3/2} - 6\sqrt{3-t} + \frac{16}{3}\right) = \frac{16}{3}.$

40. Converges. $\displaystyle\int_0^2 \frac{dx}{\sqrt{4-x^2}} = \lim_{t\to 2^-}\arcsin(t/2) = \pi/2.$

41. Diverges. $\displaystyle\int_1^\infty \frac{dx}{x(\ln x)^2} = \int_1^2 \frac{dx}{x(\ln x)^2} + \int_2^\infty \frac{dx}{x(\ln x)^2}$

$$= \lim_{s\to 1^+}\left(\frac{1}{\ln s} - \frac{1}{\ln 2}\right) + \lim_{t\to\infty}\left(\frac{1}{\ln 2} - \frac{1}{\ln t}\right) = \infty.$$

42. Diverges. $\displaystyle\int_0^\infty \frac{dx}{(x-1)^2} = \int_0^1 \frac{dx}{(x-1)^2} + \int_1^2 \frac{dx}{(x-1)^2} + \int_2^\infty \frac{dx}{(x-1)^2} = \infty.$

43. Diverges. $\displaystyle\int_0^\infty \frac{dx}{e^x-1} = \int_0^1 \frac{dx}{e^x-1} + \int_1^\infty \frac{dx}{e^x-1} = \infty.$

44. Diverges. $\displaystyle\int_{-\infty}^\infty e^{-x}\,dx = \int_{-\infty}^0 e^{-x}\,dx + \int_0^\infty e^{-x}\,dx = \infty.$

45. Converges. $\displaystyle\int_{-\infty}^\infty \frac{dx}{e^x+e^{-x}} = \int_{-\infty}^\infty \frac{e^x}{e^{2x}+1}\,dx = \frac{\pi}{2}.$

46. Converges. $\displaystyle\int_0^1 \frac{e^{-\sqrt{x}}}{\sqrt{x}}\,dx = 2-2/e.$

47. Converges. $\displaystyle\int_0^{\pi/2} \frac{\cos x}{\sqrt{\sin x}}\,dx = 2.$

48. Diverges. $\displaystyle\int_0^{\pi/2} \sec^2 x\,dx = \lim_{t\to\pi/2^-}\tan t = \infty.$

49.   (a) If $p>1$, $\displaystyle\int_1^\infty \frac{dx}{x^p} = \lim_{t\to\infty}\frac{1-t^{1-p}}{p-1} = \frac{1}{p-1}.$

(b) If $p < 1$, $\displaystyle\int_1^\infty \frac{dx}{x^p} = \lim_{t\to\infty} \frac{1-t^{1-p}}{p-1} = \infty$.

50. $\displaystyle\int_0^1 \frac{dx}{x^p} = \lim_{t\to 0^+}\int_t^1 \frac{dx}{x^p} = \lim_{t\to 0^+}\frac{1-t^{1-p}}{p-1}$. This limit is a finite number only if $p < 1$. Thus, $\displaystyle\int_0^1 \frac{dx}{x^p}$ converges if $p < 1$ and diverges if $p \geq 1$.

51. Substituting $u = \ln x$ and $du = \dfrac{dx}{x}$ produces the equivalent integral $\displaystyle\int_0^1 \frac{du}{u^p}$, which converges for $p < 1$ (by problem 50).

52. Substituting $u = \ln x$ and $du = \dfrac{dx}{x}$ produces the equivalent integral $\displaystyle\int_1^\infty \frac{du}{u^p}$, which converges for $p > 1$ (by problem 49).

53. The integral diverges for all p. This is quite clear if $p \geq 0$, but it's true for negative p as well. One reason is that $x^p e^x \to \infty$ as $x \to \infty$ for all p.

54. The integral converges for all p.

    This is clear if $p \leq 0$, but it's true for positive p as well. A rigorous proof takes some effort.

55. (a) The integrand is unbounded near $x = 0$.

    (b) l'Hôpital's rule gives $\displaystyle\lim_{x\to 0^+}\frac{\sin x}{x} = 1$, so the integrand is actually bounded on the whole interval of integration.

56. l'Hôpital's rule implies that $\displaystyle\lim_{x\to 0^+} f(x) = 0$, so f is continuous on all of $[0, 1]$, and the integral is proper.

57. Note that $f(x) = x/\ln x \to 0$ as $x \to 0^+$. Thus f is bounded, so the integral is proper.

58. $\displaystyle\int_1^\infty f'(x)\,dx = \lim_{t\to\infty} f(t) - f(1)$. Thus, the improper integral converges if (and only if) $\displaystyle\lim_{t\to\infty} f(t)$ is a finite number. Since $|f(x)| \leq e^{-x}\ln x$, $\displaystyle\lim_{t\to\infty}|f(t)| \leq \lim_{t\to\infty} e^{-t}\ln t = 0$. Thus, $\displaystyle\lim_{t\to\infty} f(t) = 0$ (and so $\displaystyle\int_1^\infty f'(x)\,dx = -f(1)$).

59. $\displaystyle\int_0^\infty \left(\frac{2x}{x^2+1} - \frac{C}{2x+1}\right)dx = \ln(x^2+1) - \frac{C}{2}\ln(2x+1) = \ln\left(\frac{x^2+1}{\sqrt{(2x+1)^C}}\right)$. Thus, the improper integral converges to $-2\ln 2$ if $C = 4$ and diverges for all other values of C.

60. $\displaystyle\int_1^\infty \left(\frac{Cx}{x^2+1} - \frac{1}{2x}\right)dx = \frac{C}{2}\ln\left(x^2+1\right) - \frac{1}{2}\ln x$. Thus, the improper integral converges to $-(\ln 2)/4$ if $C = 1/2$ and diverges for all other values of C.

---

61. $\displaystyle\int_1^\infty \frac{x}{x^3+1}\,dx = \lim_{t\to\infty}\int_1^t \frac{x}{x^3+1}\,dx = \lim_{t\to\infty}-\int_1^{1/t}\frac{du}{1+u^3} = \int_0^1 \frac{du}{1+u^3}.$

62. $\displaystyle\int_0^{\pi/2}\frac{\cos x}{\sqrt{\pi-2x}}\,dx = \lim_{t\to\pi/2^-}\int_0^t\frac{\cos x}{\sqrt{\pi-2x}}\,dx = \lim_{t\to\pi/2^-}-\int_{\sqrt{\pi}}^{\sqrt{\pi-2t}}\cos\left(\frac12\left(\pi-u^2\right)\right)du$

$$= \int_0^{\sqrt{\pi}}\cos\left(\frac12\left(\pi-u^2\right)\right)du.$$

63. Substituting $u = 1/x$, $du = -dx/x^2$, and $dx = -du/u^2$ (in the integrand and in the limits of integration) gives $\displaystyle\int_1^0 -\frac{du}{u^2(1+1/u^4)} = \int_0^1\frac{u^2}{u^4+1}\,du.$

64. Let $u = e^{-x}$ and $du = -e^{-x}\,dx$. Then

$$\int_0^\infty x^3e^{-x}\,dx = \lim_{t\to\infty}\int_0^t x^3e^{-x}\,dx = \lim_{t\to 0^+}-\int_1^t(-\ln u)^3\,du$$

$$= \int_0^1(-\ln x)^3\,dx = -\int_0^1(\ln x)^3\,dx.$$

65. Using the substitution $u = 1/x$, $\displaystyle\int_1^\infty\frac{xe^{-1/x}}{1+x^4}\,dx = -\int_1^0\frac{ue^{-u}}{1+u^4}\,du = \int_0^1\frac{ue^{-u}}{1+u^4}\,du.$

    Alternatively, using the substitution $u = -1/x$, $\displaystyle\int_1^\infty\frac{xe^{-1/x}}{1+x^4}\,dx = -\int_{-1}^0\frac{ue^u}{1+u^4}\,du.$

66. $\tan x = u^{-2} \implies x = \arctan(u^{-2})$ so $dx = -\dfrac{2u^{-3}}{1+(u^{-2})^2}\,du.$ Thus,

$$\int\sqrt{1+\tan x}\,dx = \int\sqrt{1+u^{-2}}\left(-\frac{2u^{-3}}{1+u^{-4}}\,du\right) = -2\int\frac{\sqrt{u^2+1}}{u^4+1}\,du.$$

    Because $\tan(\pi/4) = 1 \implies u = 1$ and $x\to\pi/2^- \implies u\to 0$, we have

$$\int_{\pi/4}^{\pi/2}\sqrt{1+\tan x}\,dx = -2\int_1^0\frac{\sqrt{u^2+1}}{1+u^4}\,du = 2\int_0^1\frac{\sqrt{u^2+1}}{1+u^4}\,du.$$

67. Let $u = 1/x$. Then $x = 1/u$ and $dx = -du/u^2$, and so

$$\int_0^1\frac{dx}{\sqrt{x^4}\sqrt{x^{-1}+x^{-3}}} = \lim_{t\to 0^+}\int_t^1\frac{dx}{x^2\sqrt{x^{-1}+x^{-3}}} \to \lim_{t\to 0^+}-\int_{1/t}^1\frac{du}{u^2\sqrt{u^{-1}+u^{-3}}}$$

$$= \lim_{t\to 0^+}\int_1^{1/t}\frac{du}{\sqrt{u^3+u}}$$

$$= \int_1^\infty\frac{du}{\sqrt{u^3+u}} = \int_1^\infty\frac{dx}{\sqrt{x+x^3}}.$$

§10.2   Detecting Convergence, Estimating Limits

1. (a) $\dfrac{1}{g(x)} < \dfrac{1}{f(x)} < 1.$

    (b) $\dfrac{1}{(g(x))^r} < \dfrac{1}{(f(x))^r} < 1$; note that this holds for any $r \geq 1$.

    (c) $\dfrac{1}{(g(x))^r} < \dfrac{1}{(f(x))^r} < 1$, just as in (b).

2. (a) $1 < \dfrac{1}{g(x)} < \dfrac{1}{f(x)}$

    (b) For any $r \geq 1$, $1 < \dfrac{1}{\big(g(x)\big)^r} < \dfrac{1}{\big(f(x)\big)^r}.$

    (c) As in part (b), if $r > 0$, then $1 < \dfrac{1}{\big(g(x)\big)^r} < \dfrac{1}{\big(f(x)\big)^r}.$

3. (a) $f(x) = \dfrac{x^2+2}{x^2+1} = \dfrac{x^2+1+1}{x^2+1} = 1 + \dfrac{1}{x^2+1} > 1.$

    (b) The inequality in part (a) shows that $\displaystyle\int_{-\infty}^{\infty} f(x)\,dx > \int_{-\infty}^{\infty} 1\,dx$; the second integral is easily seen to diverge to infinity.

4. (a) If $x \geq 0$ then $e^x \geq 1$, so $f(x) = \dfrac{e^x}{1+e^x} > \dfrac{e^x}{e^x+e^x} = \dfrac{1}{2}.$

    (b) For $\displaystyle\int_{-\infty}^{\infty} f(x)\,dx$ to converge requires that $\displaystyle\int_{0}^{\infty} f(x)\,dx$ must converge. But the inequality in part (a) shows $\displaystyle\int_{0}^{\infty} f(x)\,dx > \int_{0}^{\infty} \dfrac{1}{2}\,dx$, and the last integral is easily shown to diverge to infinity.

5. (a) For every real number x, we have $-1 \leq \sin x \leq 1$, which implies that $x - 1 \leq f(x) \leq x + 1$ for all x.

    (b) The inequality $x - 1 \leq f(x)$ implies that $\displaystyle\int_{2}^{\infty} (x-1)\,dx \leq \int_{2}^{\infty} (x + \sin x)\,dx$. The left-hand integral is easily shown to diverge to infinity, and so must the right-hand integral.

6. (a) For every real number x, we have $-1 \leq \sin x \leq 1$, and so $x^2 - 1 \leq f(x) \leq x^2 + 1$ for all x.

    (b) The inequality $x^2 - 1 \leq f(x)$ implies that $\displaystyle\int_{2}^{\infty} (x^2 - 1)\,dx \leq \int_{2}^{\infty} (x^2 + \sin x)\,dx.$ The left-hand integral is easily shown to diverge to infinity, and so the right-hand integral also diverges to infinity.

7. (a) $0 \leq \sqrt{x}$ for all $x \geq 0$, so $x^2 \leq x^2 + \sqrt{x} = f(x)$ for all $x \geq 0$.

(b) Part (a) implies that $\int_1^\infty \dfrac{dx}{f(x)} \le \int_1^\infty \dfrac{dx}{x^2} = 1$; it follows that $\int_1^\infty \dfrac{dx}{f(x)}$ converges (to a value less than 1).

8. (a) $0 \le \sqrt{x} \le x^2/2$ for all $x \ge 2$, so $x^2/2 = x^2 - x^2/2 \le x^2 - \sqrt{x} \le x^2$ for all $x \ge 2$.

   (b) Converges; $0 \le \int_3^\infty \dfrac{dx}{x^2 - \sqrt{x}} \le 2 \int_3^\infty \dfrac{dx}{x^2} = \dfrac{2}{3}$.

9. Diverges; $\int_2^\infty \dfrac{dx}{x - \sqrt{x}} > \int_2^\infty \dfrac{dx}{x}$, and the right-hand integral diverges.

10. Diverges. Since $\sqrt{x} \le x$ for all $x \ge 1$, we have $x + \sqrt{x} \le 2x$ for $x \ge 1$, and so $\int_1^\infty \dfrac{dx}{x + \sqrt{x}} > \int_1^\infty \dfrac{dx}{2x}$; the last integral is easily seen to diverge.

11. $\int_a^\infty e^{-x}\, dx = e^{-a} \le 10^{-5}$ if $a \ge \ln 100000 \approx 11.513$.

12. $\int_a^\infty e^{-x}\, dx = \dfrac{e^{-a^2}}{2}$. Solving $\dfrac{e^{-a^2}}{2} = 10^{-5}$ for a gives $a = \sqrt{\ln(50000)} \approx 3.29$.

13. $\int_a^\infty \dfrac{dx}{x^2 + 1} = \dfrac{\pi}{2} - \arctan a \le 10^{-5}$ if $a \ge \tan(\tfrac{\pi}{2} - 10^{-5}) \approx 100,000$.

14. $\int_a^\infty \dfrac{dx}{x\,(\ln x)^3} = \dfrac{1}{2\,(\ln a)^2} \le 10^{-5}$ if $a \ge e^{\sqrt{50,000}} \approx 1.292 \times 10^{97}$.

15. $0 < \int_a^\infty \dfrac{dx}{x^2 + e^x} < \int_a^\infty e^{-x}\, dx = e^{-a} \le 10^{-5}$ if $a \ge \ln(100000) \approx 11.513$. Thus, $\int_0^{12} \dfrac{dx}{x^2 + e^x}$ approximates $\int_0^\infty \dfrac{dx}{x^2 + e^x}$ within 10^{-5}.

16. $0 < \int_a^\infty \dfrac{dx}{x^4\sqrt{2x^3 + 1}} < \dfrac{1}{\sqrt{2}} \int_a^\infty \dfrac{dx}{x^{11/2}} = \dfrac{\sqrt{2}}{9} a^{-9/2} \le 10^{-5}$ if $a \ge \left(\dfrac{2 \times 10^{10}}{81}\right)^{1/9} \approx 8.5606$. Thus, $\int_1^9 \dfrac{dx}{x^4\sqrt{2x^3 + 1}}$ approximates $\int_1^\infty \dfrac{dx}{x^4\sqrt{2x^3 + 1}}$ within 10^{-5}.

17. $0 < \int_a^\infty \dfrac{\arctan x}{\left(1 + x^2\right)^3}\, dx < \dfrac{\pi}{2} \int_a^\infty \dfrac{dx}{x^6} = \dfrac{\pi}{10a^5} \le 10^{-5}$ if $a \ge \left(10^4\pi\right)^{1/5} \approx 7.9329$. Thus, $\int_0^8 \dfrac{\arctan x}{\left(1 + x^2\right)^3}\, dx$ approximates $\int_0^\infty \dfrac{\arctan x}{\left(1 + x^2\right)^3}\, dx$ within 10^{-5}.

18. $0 < \int_a^\infty \dfrac{e^{-x}}{2 + \cos x}\, dx < \int_a^\infty e^{-x}\, dx = e^{-a} \le 10^{-5}$ if $a \ge \ln(100000) \approx 11.513$. Thus, $\int_0^{12} \dfrac{e^{-x}}{2 + \cos x}\, dx$ approximates $\int_0^\infty \dfrac{e^{-x}}{2 + \cos x}\, dx$ within 10^{-5}.

19. Let $I = \int_1^\infty f(x)\,dx$. Then, if $a > 1$, $I - \int_1^a f(x)\,dx = \int_a^\infty f(x)\,dx$. Since $\lim\limits_{a\to\infty} \int_1^a f(x)\,dx = I$, $\lim\limits_{a\to\infty} \int_a^\infty f(x)\,dx = 0$. This implies that if a is large enough, $\int_a^\infty f(x)\,dx \leq 10^{-10}$.

20. Since $g(x) \geq 0$ for all $x \geq 1$, $\int_1^b g(x)\,dx$ is an increasing function of b. Since $\lim\limits_{t\to\infty} \int_1^t g(x)\,dx = \infty$, one can make $\int_1^b g(x)\,dx$ as large as desired by choosing b large enough.

21. $I = \int_0^\infty f(x)\,dx = \int_0^a f(x)\,dx + \int_a^\infty f(x)\,dx \implies \left| I - \int_0^a f(x)\,dx \right|$
    $= \left| \int_a^\infty f(x)\,dx \right| \leq 0.0001$.

22. $\int_0^\infty \sin\left(e^{-x}\right) dx = \lim\limits_{t\to\infty} \int_0^t \sin\left(e^{-x}\right) dx = \lim\limits_{t\to\infty} -\int_1^{e^{-t}} \sin u\, \frac{du}{u} = \int_0^1 \frac{\sin u}{u}\, du$. The desired inequalities follow from the fact that $0.8 < \dfrac{\sin x}{x} < 1$ if $0 < x < 1$.

23. $\int_0^\infty e^{-x^2}\,dx = \int_0^1 e^{-x^2}\,dx + \int_1^\infty e^{-x^2}\,dx$. Since the first integral is proper and the second is known to converge, the original improper integral must also converge.

24. (a) Let $g(x) = \sqrt{x} - \ln x$. Then $g'(x) = \dfrac{1}{2\sqrt{x}} - \dfrac{1}{x} = \dfrac{\sqrt{x} - 2}{2x}$ so the minimum value of g over the interval $(0, \infty)$ occurs at $x = 4$. Since $g(4) > 0$, $\ln x \leq \sqrt{x}$ for all $x > 0$.

    (b) The inequality in part (a) implies that $(\ln x)^2 \leq x$ for all $x \geq 2$. Therefore,
    $I = \int_2^\infty \dfrac{dx}{(\ln x)^2} \geq \int_2^\infty \dfrac{dx}{x}$. Because the improper integral on the right diverges, I also diverges.

25. The hint implies that $(\ln x)^p < x$ for large x. This implies that the given integral is comparable to $\int_2^\infty \dfrac{dx}{x}$, which diverges. Therefore I diverges for *all* positive constants p.

26. (a) Let $f(x) = \sin x - x/2$. Then $f(0) = 0$ and $f'(x) = \cos x - 1/2 > 0$ if $0 \leq x \leq 1$. Therefore, $f(x) > 0$ if $0 \leq x \leq 1$. Thus, $0 \leq \frac{1}{2}x \leq \sin x$ if $0 \leq x \leq 1$.

    (b) $0 \leq \int_0^1 \dfrac{dx}{\sqrt{\sin x}} \leq \sqrt{2} \int_0^1 \dfrac{dx}{\sqrt{x}} = 2\sqrt{2}$.

27. Diverges: Since $x^4 \leq x$ if $0 \leq x \leq 1$,
    $\int_0^\infty \dfrac{dx}{x^4 + x}\,dx = \int_0^1 \dfrac{dx}{x^4 + x}\,dx + \int_1^\infty \dfrac{dx}{x^4 + x}\,dx \geq \int_0^1 \dfrac{dx}{2x}\,dx + \int_1^\infty \dfrac{dx}{x^4 + x}\,dx$. Since

$\int_0^1 \dfrac{dx}{2x} \, dx$ diverges, the original improper integral also diverges.

28. Diverges: $\displaystyle\int_1^\infty \dfrac{dx}{\sqrt{x}-1} = \int_1^2 \dfrac{dx}{\sqrt{x}-1} + \int_2^\infty \dfrac{dx}{\sqrt{x}-1}$. Since

$\displaystyle\int_2^\infty \dfrac{dx}{\sqrt{x}-1} > \int_2^\infty \dfrac{dx}{\sqrt{x}} = \infty$, the original improper integral diverges.

29. Converges: $0 \le \displaystyle\int_1^\infty \dfrac{dx}{\sqrt[3]{x^6+x}} \le \int_1^\infty \dfrac{dx}{x^2} = 1.$

30. Converges: $0 \le \displaystyle\int_0^\infty \dfrac{dx}{\sqrt{x}(1+x)} = \int_0^1 \dfrac{dx}{\sqrt{x}(1+x)} + \int_1^\infty \dfrac{dx}{\sqrt{x}(1+x)}$

$$\le \int_0^1 \dfrac{dx}{\sqrt{x}} + \int_1^\infty \dfrac{dx}{x^{3/2}} = 2+2 = 4.$$

31. $0 < \displaystyle\int_0^\infty \dfrac{e^{-x}}{\sqrt{x}} \, dx = \int_0^1 \dfrac{e^{-x}}{\sqrt{x}} \, dx + \int_1^\infty \dfrac{e^{-x}}{\sqrt{x}} \, dx \le \int_0^1 \dfrac{dx}{\sqrt{x}} + \int_1^\infty e^{-x} \, dx = 2 + e^{-1}.$ Both of the last two summands are convergent integrals, so the original improper integral converges.

32. For $0 \le x \le 1$, $\cos x \ge \cos 1 \approx 0.54$, so $\displaystyle\int_0^1 \dfrac{\cos x}{x} \, dx \ge \int_0^1 \dfrac{0.54}{x} \, dx$, which diverges.

33.   (a) Since $\left| \dfrac{\cos x}{x^2} \right| \le \dfrac{1}{x^2}$ and $\displaystyle\int_1^\infty x^{-2} \, dx$ converges, Theorem 2 says that $\displaystyle\int_1^\infty \dfrac{\cos x}{x^2} \, dx$ converges.

   (b) Let $u = x^{-1}$ and $dv = \sin x \, dx$. Then,

$$\int_1^\infty \dfrac{\sin x}{x} \, dx = -\dfrac{\cos x}{x} \Big]_1^\infty - \int_1^\infty \dfrac{\cos x}{x^2} \, dx = \cos 1 - \int_1^\infty \dfrac{\cos x}{x^2} \, dx. \text{ Thus,}$$

$\displaystyle\int_1^\infty \dfrac{\sin x}{x} \, dx$ can be written as the difference of two finite numbers.

34. Note that I is doubly improper—the interval of integration is unbounded and the integrand is unbounded as $x \to 0^+$.

Let $I_1 = \displaystyle\int_0^1 \dfrac{dx}{\sqrt{x+x^4}}$ and let $I_2 = \displaystyle\int_1^\infty \dfrac{dx}{\sqrt{x+x^4}}$. Then, since $0 \le 1/\sqrt{x+x^4} \le 1/\sqrt{x}$ if $0 < x \le 1$, $0 \le I_1 \le \int_0^1 dx/\sqrt{x} = 2$. Furthermore, since $0 \le 1/\sqrt{x+x^4} \le 1/\sqrt{x^4}$ for all $x \ge 1$, $0 \le I_2 \le \int_1^\infty dx/\sqrt{x^4} = 1$. Therefore, $0 \le I = I_1 + I_2 \le 2 + 1 = 3.$

35. By l'Hôpital's rule, $\lim\limits_{x \to \infty} \dfrac{\int_1^x \sqrt{1 + e^{-3t}}\, dt}{x} = \lim\limits_{x \to \infty} \dfrac{\sqrt{1 + e^{-3x}}}{1} = 1$. [NOTE: The improper integral in the numerator diverges to ∞ because $\sqrt{1 + e^{-3t}} > 1$ for all $t \geq 1$. This assures that l'Hôpital's rule does apply.]

36. $\lim\limits_{x \to \infty} e^{x^2} \displaystyle\int_0^x e^{-t^2}\, dt = \infty$ because $\lim\limits_{x \to \infty} \displaystyle\int_0^x e^{-t^2}\, dt = \sqrt{\pi}/2$ and $\lim\limits_{x \to \infty} e^{x^2} = \infty$.

§10.3   Improper Integrals and Probability

2. (a) The probability density function for X is $f(x) = \dfrac{1}{\sqrt{2\pi}}e^{-x^2/2}$. The area under this function over the interval $[-3, 3]$ is the fraction of X-values which lie no more than 3 standard deviations below the mean and no more than 3 standard deviations above the mean.

   (b) $\dfrac{1}{2 \cdot \sqrt{2\pi}} \displaystyle\int_{-6}^{6} e^{-x^2/8}\, dx.$

   (c) If $u = x/2$, $du = \frac{1}{2}\, dx$. Therefore, $\dfrac{1}{2 \cdot \sqrt{2\pi}} \displaystyle\int_{-6}^{6} e^{-x^2/8}\, dx = \dfrac{1}{\sqrt{2\pi}} \displaystyle\int_{-3}^{3} e^{-u^2/2}\, du.$

   (d) Using Simpson's rule with $n = 20$, $\dfrac{1}{\sqrt{2\pi}} \displaystyle\int_{-3}^{3} e^{-x^2/2}\, dx \approx 0.9973.$

3. When $Z = \dfrac{x - 500}{100}$, $dZ = \dfrac{dx}{100}$. Furthermore, if $x = 500$, $Z = 0$; if $x = 700$, $Z = 2$. Thus,

$$I_1 = \frac{1}{100\sqrt{2\pi}} \int_{500}^{700} \exp\left(-\frac{(x - 500)^2}{2 \cdot 100^2}\right) dx$$

$$= \frac{1}{100\sqrt{2\pi}} \int_{0}^{2} \exp\left(-\frac{Z^2}{2}\right) 100 dZ = \frac{1}{\sqrt{2\pi}} \int_{0}^{2} e^{-Z^2/2}\, dZ.$$

4. Since $z = (x - m)/s$, $dz = dx/s$ and

$$\frac{1}{\sqrt{2\pi}\, s} \int \exp\left(-\frac{(x - m)^2}{2s^2}\right) dx = \frac{1}{\sqrt{2\pi}} \int e^{-z^2/2}\, dz.$$

   Using the rules for substitution in a definite integral, the limits of integration change from x_1 and x_2 to z_1 and z_2, respectively.

5. (a) Since the integrand is an even function

$$\int_{-\infty}^{\infty} e^{-x^2}\, dx = 2 \int_{0}^{\infty} e^{-x^2}\, dx = 2 \cdot \frac{\sqrt{\pi}}{2} = \sqrt{\pi}.$$

   (b) Using the substitution $u = x/\sqrt{2}$, $\dfrac{1}{\sqrt{2\pi}} \displaystyle\int_{-\infty}^{\infty} e^{\frac{-x^2}{2}} = \dfrac{1}{\sqrt{\pi}} \displaystyle\int_{-\infty}^{\infty} e^{-u^2}\, du = 1.$

   (c) Using the substitution $u = (x - m)/s$,
$$\frac{1}{\sqrt{2\pi}\, s} \int_{-\infty}^{\infty} e^{\frac{-(x-m)^2}{2s^2}} = \frac{1}{2\sqrt{\pi}} \int_{-\infty}^{\infty} e^{-u^2/2}\, du = 1.$$

6. (a) No. Negative Z-scores correspond to data values that are smaller than the mean since $z = (x - m)/s$.

   (b) 1.3 — The Z-score measures how far an observed value is from the mean in units of standard deviations.

7. (a) $z = \dfrac{600 - 500}{100} = 1.$

    (b) $z = \dfrac{450 - 500}{100} = -0.5.$

    (c) $\dfrac{1}{\sqrt{2\pi} \cdot 100} \displaystyle\int_{450}^{\infty} \exp\left(-\dfrac{(x - 500)^2}{2 \cdot 100^2}\right) dx = \dfrac{1}{\sqrt{2\pi}} \displaystyle\int_{-0.5}^{\infty} e^{-x^2/2} \, dx.$

    (d) $\dfrac{1}{\sqrt{2\pi} \cdot 100} \displaystyle\int_{-\infty}^{600} \exp\left(-\dfrac{(x - 500)^2}{2 \cdot 100^2}\right) dx = \dfrac{1}{\sqrt{2\pi}} \displaystyle\int_{-\infty}^{1} e^{-x^2/2} \, dx.$

    (e) $\dfrac{1}{\sqrt{2\pi} \cdot 100} \displaystyle\int_{450}^{600} \exp\left(-\dfrac{(x - 500)^2}{2 \cdot 100^2}\right) dx = \dfrac{1}{\sqrt{2\pi}} \displaystyle\int_{-0.5}^{1} e^{-x^2/2} \, dx.$

8. $A(z_2) - A(z_1) = \displaystyle\int_{-\infty}^{z_2} n(t) \, dt - \int_{-\infty}^{z_1} n(t) \, dt = \int_{z_1}^{z_2} n(t) \, dt.$

9. (a) Since $n(-t) = n(t),$

    $$A(-z) = \int_{-\infty}^{-z} n(t) \, dt = \int_{z}^{\infty} n(t) \, dt = \int_{-\infty}^{\infty} n(t) \, dt - \int_{-\infty}^{z} n(t) \, dt = 1 - A(z).$$

    (b) $A(-1.2) = 1 - A(1.2) \approx 1 - 0.8849 = 0.1151.$

10. Raw scores of 350, 500, and 600 correspond to Z-scores of $-1.5, 0,$ and 1, respectively. Looking at the table (and the graph above it) shows: The probability that $Z \geq 1$ is $1 - A(1) \approx 1 - 0.8413 = 0.1587.$ Thus, nearly 16% of scores are above 600.

11. The probability that $-1.5 \leq Z \leq 0$ is the same as the probability that $0 \leq Z \leq 1.5$, which is $A(1.5) - A(0) \approx 0.9333 - 0.5 = 0.4333.$ Thus about 43% of scores fall in this range.

12. In both parts $m = 10$ and $s = 5$. $z = (14 - 10)/5 = 0.8$ and $A(0.8) \approx 0.7881.$ Thus,

    $$\frac{1}{5\sqrt{2\pi}} \int_{-\infty}^{14} \exp\left(-\frac{(x - 10)^2}{50}\right) \approx 0.7881.$$

13. $z = (4 - 10)/5 = -1.2$ and $A(-1.2) = 1 - A(1.2) \approx 0.1151.$ Thus,

    $$\frac{1}{5\sqrt{2\pi}} \int_{-\infty}^{4} \exp\left(-\frac{(x - 10)^2}{50}\right) \approx 0.1151.$$

14. The idea is to estimate various areas under the graph, perhaps using the fact that each grid rectangle has area 0.05. Reasonable answers are below; they correspond to probabilities that an observation falls in the given range:

    $\approx 0.2.$

15. Counting rectangles under the graph between $x = 0.5$ and $x = 1$ gives a bit more than 8 rectangles, or just over 0.4 in area.

16. Counting rectangles under the graph between $x = 0$ and $x = 1.5$ gives about 17 rectangles, or about 0.85 in area.

17. Counting rectangles under the graph to the right of $x = 1.5$ gives about 3 rectangles, or about 0.15 in area.

18. Looking at the table shows that the top 10% starts at about 1.3 standard deviations above the mean, i.e., at a raw score of 630. Similarly, the top 5% starts at about $Z = 1.7$, i.e., at a raw score of 670.

19. The Z-scores that corresponds to differing from the average by more than 1 inch are $Z = \pm 1/0.6 \approx \pm 1.67$. Thus, the probability is (approximately)
$$\int_{-\infty}^{-1.67} n(x)\,dx + \int_{1.67}^{\infty} n(x)\,dx \approx 0.095581.$$

20. (a) The nearest edge of the net should be placed 135 feet from the cannon (i.e., so that the center of the net is 150 feet from the cannon). This position maximizes the probability that the performer will land in the net.

    (b) Missing the net corresponds to a Z-score of magnitude greater than 1.5. Thus, the probability that the performer will miss the landing net is
$$\int_{-\infty}^{-1.5} n(x)\,dx + \int_{1.5}^{\infty} n(x)\,dx \approx 0.13361.$$

21. Let D be the "nominal" diameter at which the machine is set. Then the Z-score of a can top with diameter greater than 3 inches is greater than $Z_0 = (3 - D)/0.01$. No more than 10% of the can tops produced will have diameter greater than 3 inches if $\int_{Z_0}^{\infty} n(x)\,dx = 0.1$, that is if $Z_0 \approx 1.3$ (from the table). This corresponds to $D = 2.987$ inches.

22. (a) Since $A''(x) = -\dfrac{x}{\sqrt{2\pi}} e^{-x^2/2}$, the graph of A is concave down over the interval $[0.7, 0.8]$. Therefore, the graph lies *above* the secant line joining $(0.7, A(0.7))$ and $(0.8, A(0.8))$. In other words, $A(0.75) > 0.77305$.

    (b) $A(0.75) \approx A(0.7) + 0.05 \cdot A'(0.7) \approx 0.7580 + 0.05 \cdot 0.31225 = 0.7736$. Alternatively, $A(0.75 \approx A(0.8) - 0.05 \cdot A'(0.8) \approx 0.7881 - 0.05(0.28969) \approx 0.7733$.

    (c) Since A is concave down over the interval $[0.7, 0.8]$, the line tangent to the graph $y = A(x)$ at $x = 0.7$ lies above the graph at $x = 0.75$. Thus, the estimate in part (b) overestimates $A(0.75)$.

23. (a) If $\beta < 0$, then g is not a positive function. If $\beta = 0$, then $g(x)$ is undefined. Therefore, $\beta > 0$ must be true for g to be a probability distribution.

(b) If $u = (\ln x - \alpha)/\beta$, then $du = \left(1/(\beta x)\right) dx$, so

$$\int_a^b g(x)\, dx \to \int_{(\ln a - \alpha)/\beta}^{(\ln b - \alpha)/\beta} \frac{1}{\sqrt{2\pi}} e^{-u^2/2}\, du$$

$$= \int_{-\infty}^{(\ln b - \alpha)/\beta} \frac{1}{\sqrt{2\pi}} e^{-u^2/2}\, du - \int_{-\infty}^{(\ln a - \alpha)/\beta} \frac{1}{\sqrt{2\pi}} e^{-u^2/2}\, du$$

$$= A\left(\frac{\ln b - \alpha}{\beta}\right) - A\left(\frac{\ln a - \alpha}{\beta}\right).$$

24. (a) $\displaystyle\int_{-\infty}^{\infty} f(x)\, dx = \int_1^{\infty} \frac{3}{x^4}\, dx = -\frac{1}{x^3}\Big]_1^{\infty} = 1.$

    (b) $\displaystyle\int_2^{\infty} f(x)\, dx = \int_2^{\infty} \frac{3}{x^4}\, dx = -\frac{1}{x^3}\Big]_2^{\infty} = \frac{1}{8}.$

    (c) $\displaystyle\int_{-\infty}^{3/2} f(x)\, dx = \int_1^{3/2} \frac{3}{x^4}\, dx = -\frac{1}{x^3}\Big]_1^{3/2} = \frac{19}{27}.$

    (d) $\displaystyle\int_{-\infty}^{\infty} x f(x)\, dx = \int_1^{\infty} \frac{3}{x^3}\, dx = -\frac{3}{2x^2}\Big]_1^{\infty} = \frac{3}{2}.$

    (e) We wish to determine m such that $\displaystyle\int_{-\infty}^{m} f(x)\, dx = 0.5.$ Since

    $$\int_{-\infty}^{m} f(x)\, dx = 1 - 1/m^3,$$
    $$m = 2^{1/3} \approx 1.26 \text{ minutes}.$$

25. Since $\displaystyle\int_{-\infty}^{\infty} f(x)\, dx = k\pi$, f is a probability density function if $k = 1/\pi$.

26. Since $\displaystyle\int_{-\infty}^{\infty} f(x)\, dx = \int_0^1 k(1 - x^2)\, dx = 2k/3$, f is a probability density function if $k = 3/2$.

27. (a) f is a probability density function because it is a positive function and

    $$\int_{-\infty}^{\infty} f(x)\, dx = \int_0^{\infty} \lambda e^{-\lambda x}\, dx = -e^{-\lambda x}\Big]_0^{\infty} = 1.$$

    (b) $\displaystyle\int_{-\infty}^{\infty} x f(x)\, dx = \lambda \int_0^{\infty} x e^{-\lambda x}\, dx = -\lambda \frac{1 + \lambda x}{\lambda^2} e^{-\lambda x}\Big]_0^{\infty} = \frac{1}{\lambda}.$

28. (a) The probability that a device fails between time $t = 0$ and $t = T$ months is $\int_0^T f(x)\, dx$. Thus, for this device, $0.1 = \int_0^6 f(x)\, dx = 1 - e^{-6\lambda}$ so $\lambda = -\ln(0.9)/6 \approx 0.01756$.

    (b) The probability that a device is still functioning after T months is $\int_T^{\infty} f(x)\, dx$. Using the value of λ from part (a), the probability is 0.729.

(c) The mean lifespan of the device is $\int_{-\infty}^{\infty} x f(x)\,dx = -6/\ln(0.9) \approx 56.95$ months.

29. (a) $\Gamma(1) = \int_0^{\infty} e^{-t}\,dt = \lim_{a \to \infty} \int_0^a e^{-t}\,dt = \lim_{a \to \infty} \left(1 - e^{-a}\right) = 1.$

    (b) If $x > 1$, an integration by parts (with $u = t^{x-1}$ and $dv = e^{-t}\,dt$) shows that

$$\Gamma(x) = \int_0^{\infty} t^{x-1} e^{-t}\,dt = -t^{x-1} e^{-t} \Big]_0^{\infty} + (x-1) \int_0^{\infty} t^{x-2} e^{-t}\,dt = 0 + (x-1)\Gamma(x-1).$$

30. $\Gamma\left(\frac{2}{3}\right) = \int_0^{\infty} \dfrac{dt}{\sqrt[3]{t}\, e^t} = \int_0^1 \dfrac{dt}{\sqrt[3]{t}\, e^t} + \int_1^{\infty} \dfrac{dt}{\sqrt[3]{t}\, e^t} < \int_0^1 \dfrac{dt}{\sqrt[3]{t}} + \int_1^{\infty} \dfrac{dt}{e^t}$

    $= 3/2 + e^{-1} \approx 1.8679 < 2.$

    Finally, $\Gamma\left(\frac{2}{3}\right) > 0$ since the integrand $(t^{-1/3} e^{-t})$ is positive throughout the interval of integration.

31. The substitution $u = \sqrt{t}$ shows that $\Gamma\left(\frac{1}{2}\right) = \int_0^{\infty} \dfrac{dt}{\sqrt{t}\, e^t} = 2 \int_0^{\infty} e^{-u^2}\,du = \sqrt{\pi}.$

32. Let $t = \ln(1/x)$. Then $e^{-t} = x$ and $dt = -(1/x)\,dx$ so

$$\int_0^{\infty} t^z e^{-t}\,dt = \lim_{s \to \infty} \int_0^s t^z e^{-t}\,dt = \lim_{w \to 0^+} -\int_1^w \left(\ln(1/x)\right)^z dx = \int_0^1 \left(\ln(1/x)\right)^z dx.$$

33. First make the substitution $x = e^{-u}$, then make the substitution $w = (m+1)u$:

$$\int_0^1 x^m (\ln x)^n\,dx \to -\int_{\infty}^0 e^{-(m+1)u}(-u)^n\,du = (-1)^n \int_0^{\infty} u^n e^{-(m+1)u}\,du$$

$$\to \frac{(-1)^n}{(m+1)^{n+1}} \int_0^{\infty} w^n e^{-w}\,dw = \frac{(-1)^n\, n!}{(m+1)^{n+1}}.$$

34. (a) The sample space is the space of all possible "outcomes"—in this case that's the set of all points (x, y) with both x and y in the interval $[0, 2]$; this is another description for the square $[0, 2] \times [0, 2]$.

    (b) All points in the square $[0, 2] \times [0, 2]$ are equally likely. The outcomes $0 \le x \le 1$ and $0 \le y \le 1$ corresponds to the "lower left quarter" of this square, which has area $1/4$ that of whole square $[0, 2] \times [0, 2]$.

35. The entire sample space is the square $[-2, 2] \times [-2, 2]$, which has area $4^2 = 16$. The probability that $x^2 + y^2 \le 1$ is the proportion of the total area occupied by the interior of the circle $x^2 + y^2 \le 1$. This circle has area π, so the probability is $\pi/16 \approx 0.196$.

36. The sample space is the square $[0, 1] \times [0, 1]$, which has area 1. The probability that $x + y \le 1/2$ is the proportion of the total area occupied by the triangle with corners $(0, 0)$, $(0, 1/2)$, and $(1/2, 0)$. (In this triangle, $x + y \le 1/2$.) This triangle has area $1/8$; this is the desired proportion.)

37. The sample space is the square $[0, 1] \times [0, 1]$, which has area 1. The probability that $y/x \geq 2$ is the proportion of the total area occupied by the triangle with corners $(0, 0)$, $(0, 1)$, and $(1/2, 1)$. (In this triangle, $y/x \geq 2$.) This triangle has area $1/4$; this is the desired proportion.)

38. The sample space is the square $[0, 2] \times [0, 2]$, which has area 4. The probability that $xy \leq 2$ is the proportion of the total area occupied by the part of the square that lies below the curve $y = 2/x$. The complementary region (where $y \geq 2/x$) may be a bit easier to calculate by integration: It's $\int_1^2 \left(2 - \dfrac{2}{x}\right) dx = 2 - \ln 4$. Thus the desired area is $4 - (2 - \ln 4) = 2 + \ln 4$, and the desired probability is the ratio $(2 + \ln 4)/4 \approx 0.847$.

39. The sample space is the square $[0, 3] \times [0, 3]$, which has area 9. The probability that $x + y \geq xy$ is the proportion of the total area occupied by the part of the square that lies below the curve $y = x/(x - 1)$. A look at graphs shows that this part of the square has area $\dfrac{9}{2} + \int_{3/2}^3 \dfrac{x}{x - 1} dx = 6 + \ln 4 \approx 7.386$, so the desired probability is $(6 + \ln 4)/9 \approx 0.821$.

11   Infinite Series

§11.1   Sequences and Their Limits

1. $\lim\limits_{k\to\infty} a_k$ does not exist: $\lim\limits_{k\to\infty} a_{2k} = \infty$ but $\lim\limits_{k\to\infty} a_{2k+1} = -\infty$.

2. $\lim\limits_{k\to\infty} a_k = 0$.

3. $\lim\limits_{k\to\infty} a_k = \infty$.

4. $\lim\limits_{k\to\infty} a_k = 0$.

5. $\lim\limits_{k\to\infty} a_k = 0$.

6. $\lim\limits_{k\to\infty} a_k$ does not exist because the values of $\sin k$ do not approach a single value as $k \to \infty$.

7. $\lim\limits_{k\to\infty} a_k = \pi/2$ because $\lim\limits_{x\to\infty} \arctan x = \pi/2$.

8. $\lim\limits_{k\to\infty} a_k = 1$ because $\lim\limits_{k\to\infty} 1/k = 0$ and $\cos 0 = 1$.

9. $\lim\limits_{k\to\infty} a_k = 0$.

10. $\lim\limits_{k\to\infty} a_k$ does not exist because the values of a_k are alternately -1 and $+1$.

11. $\lim\limits_{k\to\infty} a_k = 1$.

12. $\lim\limits_{k\to\infty} a_k = \sqrt{2}$.

13. $\lim\limits_{m\to\infty} a_m = \lim\limits_{m\to\infty} \dfrac{m^2}{e^m} = \lim\limits_{m\to\infty} \dfrac{2m}{e^m} = \lim\limits_{m\to\infty} \dfrac{2}{e^m} = 0$.

14. $\lim\limits_{j\to\infty} a_j = \lim\limits_{j\to\infty} \dfrac{\ln j}{\sqrt[3]{j}} = \lim\limits_{j\to\infty} \dfrac{3 j^{2/3}}{j} = \lim\limits_{j\to\infty} \dfrac{2}{\sqrt[3]{j}} = 0$.

15. Since $\dfrac{k!}{(k+1)!} = \dfrac{1}{k+1}$, $\lim\limits_{k\to\infty} a_k = 0$.

16. Since $a_k = \ln\left(\dfrac{k}{k+1}\right) = \ln\left(1 - \dfrac{1}{k+1}\right)$ and $\lim\limits_{k\to\infty} \dfrac{1}{k+1} = 0$, $\lim\limits_{k\to\infty} a_k = 0$.

17. $\lim\limits_{k\to\infty} \ln a_k = \lim\limits_{k\to\infty} \dfrac{\ln 3}{k} = 0$. Therefore, $\lim\limits_{k\to\infty} a_k = 1$.

18. (a) $\lim\limits_{n\to\infty} x^n$ diverges when $x > 1$ or $x \le -1$.

    (b) $\lim\limits_{n\to\infty} x^n = 0$ when $|x| < 1$.

    (c) $\lim\limits_{n\to\infty} x^n$ converges to 1 when $x = 1$.

19. $a_k = (-1)^k/k$. $\lim\limits_{k\to\infty} a_k = 0$, but $\{a_k\}$ is neither an increasing nor a decreasing sequence.

20. $a_k = \cos k$; $|a_k| \le 1$ for all integers $k \ge 1$, but $\lim\limits_{k\to\infty} a_k$ does not exist.

21. $a_k = k$; $a_k < a_{k+1}$ for all integers $k \ge 1$, but $\lim\limits_{k\to\infty} a_k = \infty$.

22. $a_k = -k$; $a_{k+1} < a_k$ for all integers $k \ge 1$, and $\lim\limits_{k\to\infty} a_k = -\infty$.

23. $a_k = e^{-k}$; $a_{k+1} < a_k$ for all integers $k \ge 1$, and $\lim\limits_{k\to\infty} a_k = 0$.

24. $a_k = (-1)^k k$; $a_k < a_{k+1}$ when k is an odd integer, but $a_{k+1} < a_k$ when k is an even integer. Thus, $\{a_k\}$ is not a monotone sequence. Also, $\lim\limits_{k\to\infty} |a_k| = \infty$, so the sequence is not bounded.

25. The sequence is bounded above (e.g, by 0.8) and is monotone increasing and by Theorem 3 has a limit.

26. $\lim\limits_{k\to\infty} b_k = \lim\limits_{k\to\infty} (L - a_k) = \lim\limits_{k\to\infty} L - \lim\limits_{k\to\infty} a_k = L - \lim\limits_{k\to\infty} a_k = L - L = 0.$

27. $\lim\limits_{n\to\infty} a_n = \lim\limits_{n\to\infty} n \sin(1/n) = \lim\limits_{n\to\infty} \dfrac{\sin\left(n^{-1}\right)}{n^{-1}} = \lim\limits_{n\to\infty} \cos\left(n^{-1}\right) = 1.$

28. $\lim\limits_{k\to\infty} \dfrac{\ln(2^k + 3^k)}{k} = \ln 3 \implies \lim\limits_{k\to\infty} \left(2^k + 3^k\right)^{1/k} = e^{\ln 3} = 3.$

29. $\lim\limits_{k\to\infty} a_k = \displaystyle\int_0^\infty e^{-x}\, dx = 1.$

30. a_k is the "tail" of the convergent improper integral $\displaystyle\int_0^\infty \dfrac{dx}{1 + x^2}$. Thus, $\lim\limits_{k\to\infty} a_k = 0$.

31. $a_k = \sqrt{k^2 + 1} - k = \dfrac{\left(\sqrt{k^2 + 1} - k\right) \cdot \left(\sqrt{k^2 + 1} + k\right)}{\sqrt{k^2 + 1} + k} = \dfrac{1}{\sqrt{k^2 + 1} + k}$. Thus, $\lim\limits_{k\to\infty} a_k = 0$.

32. $a_k = \sqrt{k^2 + k} - k = \dfrac{\left(\sqrt{k^2 + k} - k\right) \cdot \left(\sqrt{k^2 + k} + k\right)}{\sqrt{k^2 + k} + k} = \dfrac{k}{\sqrt{k^2 + k} + k}$. Thus, $\lim\limits_{k\to\infty} a_k = 1/2$.

33. $\lim\limits_{k\to\infty} a_k$ exists only when $x \le 0$ because $e^x > 1$ when $x > 0$. When $x < 0$, $\lim\limits_{k\to\infty} a_k = 0$. When $x = 0$, $\lim\limits_{k\to\infty} a_k = 1$.

34. $\lim_{k\to\infty} a_k = 0$ when $1/e < x < e$ and $\lim_{k\to\infty} a_k = 1$ when $x = e$.

35. $\lim_{k\to\infty} a_k = 0$ when $-\sin 1 < x < \sin 1$ and $\lim_{k\to\infty} a_k = 1$ when $x = \sin 1$.

36. Since $-2 < -\pi/2 < \arctan x < \pi/2 < 2$, $\lim_{k\to\infty} a_k = 0$ when $-\infty < x < \infty$.

37. $\lim_{n\to\infty} \dfrac{\ln x}{n} = 0 \implies \lim_{n\to\infty} x^{1/n} = e^0 = 1$ for all $x > 0$.

38. (a) $\lim_{k\to\infty} \ln(a_k) = \lim_{k\to\infty} \ln (1 + x/k)^k = \lim_{k\to\infty} k \ln (1 + x/k)$

   $$= \lim_{k\to\infty} \frac{\ln (1 + x/k)}{k^{-1}} = \lim_{k\to\infty} \frac{x}{1 + x/k} = x.$$

   (b) $\lim_{k\to\infty} a_k = e^x$.

39. Let $a_n = \left(1 - \dfrac{1}{2n}\right)^n$. Then $\ln(a_n) = n \ln \left(1 - \dfrac{1}{2n}\right) = \dfrac{\ln \left(1 - \frac{1}{2n}\right)}{\frac{1}{n}}$ so, by l'Hôpital's rule,

   $\lim_{n\to\infty} \ln(a_n) = -1/2$. It follows that $\lim_{n\to\infty} a_n = e^{-1/2}$.

40. (a) $a_{10} = \sum_{k=1}^{10} \dfrac{k}{100} = 0.01 + 0.02 + \cdots + 0.10 = 0.55.$

   (b) $a_n = R_n$, the right Riemann sum approximation to $\int_0^1 x\,dx$ with n subintervals.

41. (a) $a_{n+1} = \sum_{k=1}^{n+1} \dfrac{1}{(n+1)+k} = \sum_{k=2}^{n+2} \dfrac{1}{n+k} = \sum_{k=2}^{n} \dfrac{1}{n+k} + \dfrac{1}{2n+1} + \dfrac{1}{2n+2}$

   $$> \sum_{k=2}^{n} \frac{1}{n+k} + \frac{2}{2n+2} = \sum_{k=1}^{n} \frac{1}{n+k} = a_n.$$

   (b) $a_n = \sum_{k=1}^{n} \dfrac{1}{n+k} \le \sum_{k=1}^{n} \dfrac{1}{n+1} = \dfrac{n}{n+1} < 1$ when $n \ge 1$.

   (c) Parts (a) and (b) imply that the sequence $\{a_n\}$ is monotonically increasing and bounded above. Therefore, $\lim_{n\to\infty} a_n$ exists.

   (d) Part (a) implies that the sequence $\{a_n\}$ is monotonically increasing. Therefore, $\lim_{n\to\infty} a_n > a_k$ for any integer $k \ge 1$. Since $a_1 = 1/2$, $\lim_{n\to\infty} a_n > 1/2$.

   (e) $\lim_{n\to\infty} a_n = \int_0^1 \dfrac{dx}{1+x} = \ln 2$.

42. $\lim_{n\to\infty} a_n = \int_0^1 x^2\,dx = \dfrac{1}{3}$.

43. (a) First, note that $\ln(\sqrt[n]{n!}) = \ln(n!)^{1/n} = \frac{1}{n}\ln(n!) = \frac{1}{n}\sum_{k=1}^{n}\ln k$. Also, $n = \left(n^n\right)^{1/n}$ so

$$\ln n = \frac{1}{n}\ln(n^n) = \frac{1}{n}\sum_{k=1}^{n}\ln n. \text{ Therefore,}$$

$$\ln a_n = \ln\left(\frac{\sqrt[n]{n!}}{n}\right) = \ln\left(\sqrt[n]{n!}\right) - \ln n = \frac{1}{n}\sum_{k=1}^{n}\ln k - \frac{1}{n}\sum_{k=1}^{n}\ln n.$$

(b) The right sum approximation to $\int_0^1 \ln x\, dx$ is

$$R_n = \frac{1}{n}\sum_{k=1}^{n}\ln\left(\frac{k}{n}\right) = \frac{1}{n}\sum_{k=1}^{n}\ln k - \frac{1}{n}\sum_{k=1}^{n}\ln n = \ln a_n.$$

(c) The result in part (b) implies that $\lim_{n\to\infty}\ln a_n = \int_0^1 \ln x\, dx = x\ln x - x\Big]_0^1 = -1.$
Therefore, $\lim_{n\to\infty} a_n = e^{-1}$.

44. (a) $a_{n+1} \le a_n$ for all $n \ge 3$.

(b) The sequence $\{a_n\}$ is monotonically decreasing for all $n \ge 3$ and bounded below (by zero). Therefore, the sequence must converge.

(c) Observe that $0 < a_n \le (4/5)^{n-4}(32/3)$ for all $n \ge 4$. Since $\lim_{n\to\infty}(4/5)^{n-4} = 0$, $\lim_{n\to\infty} a_n = 0$ (by the "squeeze" theorem).

45. Observe that $\frac{n+3}{2n+1} \le \frac{6}{7}$ for all $n \ge 3$. Now,

$$\frac{a_{n+1}}{a_3} = \frac{a_{n+1}}{a_n} \cdot \frac{a_n}{a_{n-1}} \cdots \frac{a_4}{a_3} \le \left(\frac{6}{7}\right)^{n-2} \implies a_{n+1} \le a_3(6/7)^{n-2} \implies \lim_{n\to\infty} a_n = 0.$$

46. Yes. $|a_n| \le 1$ since $|\cos x| \le 1$ for all x.

47. No. The terms of the sequence change sign.

48. Yes. $|a_n| \le 1$.

49. Yes. $|a_{n+1}| < |a_n|$

50. (a) $a_1 = 1/2$; $a_2 = 3/8$; $a_5 = 945/3840 = 63/256$

(b) Since $0 < a_{n+1} < a_n$, the sequence is bounded below and monotonically decreasing. Therefore, $\lim_{n\to\infty} a_n$ exists.

51. Let $a_n = \sin\left(\pi/n^2\right)$. Then, the inequality $0 < \sin x < x$ when $0 < x < 1$ implies that $0 < a_n < \pi/n^2$. Since $\lim_{n\to\infty} \pi/n^2 = 0$, Theorem 2 implies the result.

52. No. The terms of the sequence alternate between 1 and 0.

53. Yes. The inequalities $0 < a_{n+1} = a_n/2 < a_n$ imply that the sequence is monotonically decreasing and bounded below.

54. (a) Since $0 < a_{n+1} < a_n$, the sequence is bounded below and monotone decreasing. Therefore, it must have a limit.

    (b) $\lim_{n \to \infty} a_n = \lim_{n \to \infty} 1/n = 0$.

55. $a_{n+1} = x^{1/2^n}$ so $\lim_{n \to \infty} a_n = 1$ for all $x > 0$ by the previous problem. When $x = 0$, $\lim_{n \to \infty} a_n = 0$. Thus, $\lim_{n \to \infty} a_n$ exists for all $x \geq 0$.

§11.2 Infinite Series, Convergence, and Divergence

1. (a) $a_1 = 1/5, a_2 = 1/25, a_5 = 1/3125, a_{10} = 1/9765625.$

$S_n = \sum_{k=0}^{n} \frac{1}{5^k} = \frac{1 - (1/5)^{n+1}}{1 - (1/5)} = \frac{5 - (1/5)^n}{4}.$ Thus, $S_1 = 6/5 = 1.2,$
$S_2 = 31/25 = 1.24, S_5 = 3906/3125 = 1.24992,$ and
$S_{10} = 12207031/9765625 = 1.2499999744.$

   (b) For any $k \geq 0, 0 < a_{k+1} = \frac{1}{5^{k+1}} = \frac{1}{5}\frac{1}{5^k} = \frac{a_k}{5} < a_k.$ Thus, the sequence $\{a_k\}$ is decreasing and bounded below.

   (c) $S_{n+1} = S_n + 1/5^{n+1} > S_n,$ so the sequence is increasing. Since $S_n = \frac{5}{4} - \frac{1}{4}\left(\frac{1}{5}\right)^n,$
   $S_n < 5/4$ for all $n \geq 0.$ Because the sequence of partial sums is increasing and bounded above, it must converge.

   (d) $S_n = \frac{5}{4} - \frac{1}{4}\left(\frac{1}{5}\right)^n \implies \lim_{n\to\infty} S_n = \frac{5}{4}.$

   (e) $R_1 = 1/20, R_2 = 1/100, R_5 = 1/12500, R_{10} = 1/39062500.$

   (f) $R_n = \sum_{k=n+1}^{\infty} 5^{-k} = \frac{1}{4}\left(\frac{1}{5}\right)^n.$ Thus, $0 < R_{n+1} < R_n$ for all $n \geq 0.$

   (g) $\lim_{n\to\infty} R_n = 0.$

2. (a) $a_1 = -4/5 = -0.8; a_2 = 16/25 = 0.64; a_5 = -1024/3125 = -0.32768;$
   $a_{10} = 1048576/9765625 = 0.1073741824.$
   $S_1 = 1/5 = 0.2; S_2 = 21/25 = 0.84; S_5 = 1281/3125 = 0.40992;$
   $S_{10} = 5891381/9765625 \approx 0.60328.$

   (b) $\sum_{k=0}^{\infty}(-0.8)^k = \frac{1}{1 - (-0.8)} = \frac{5}{9} \approx 0.5555555555.$

   (c) $R_1 = 16/45 \approx 0.35556; R_2 = -64/225 \approx -0.28444; R_5 = 4096/28125 \approx 0.14564;$
   $R_{10} = -4194304/87890625 \approx -0.047722.$

   (d) The sequence is not decreasing ($a_{2k-1} < a_{2k+1} < a_{2k}$, but it is bounded ($|a_k| \leq 1$).

   (e) No — $S_{2m+1} < S_{2m} < S_{2m-1}$ for $m = 1, 2, 3, \ldots$ (the partial sums go up and down in value).

   (f) $R_n = \sum_{k=n+1}^{\infty}(-0.8)^k = \frac{5}{9}(-0.8)^{n+1}$ so the terms of the sequence alternate between positive and negative.

   (g) $|R_{n+1}| = \left|\sum_{k=n+2}^{\infty}(-0.8)^k\right| = \left|-0.8\sum_{k=n+1}^{\infty}(-0.8)^k\right| = 0.8|R_n| < |R_n|$ (i.e., the sequence is decreasing).

(h) $\lim\limits_{n \to \infty} R_n = 0$.

3. (a) $a_{k+1} = \dfrac{1}{(k+1) + 2^{k+1}} < \dfrac{1}{k + 2^k} = a_k$ since $k < k+1$ and $2^k < 2^{k+1}$.

    (b) If $k \geq 0$, $a_k = \dfrac{1}{k + 2^k} \leq \dfrac{1}{2^k} = 2^{-k}$.

    (c) $S_{n+1} = S_n + \dfrac{1}{(n+1) + 2^{n+1}} > S_n$ for all $n \geq 0$.

    (d) $S_n = \sum\limits_{k=0}^{n} a_k \leq \sum\limits_{k=0}^{n} 2^{-k} = \dfrac{1 - \left(\frac{1}{2}\right)^{n+1}}{1 - \frac{1}{2}} = 2 - 2^{-n} < 2$ for all $n \geq 0$.

    (e) From parts (c) and (d) we know that the sequence of partial sums is increasing and bounded above. Therefore, the sequence of partial sums converges (i.e., the series converges).

4. (a) $S_1 = 8/15 \approx 0.53333$; $S_2 = 103/165 \approx 0.62424$; $S_5 = 2626616/3892119 \approx 0.67486$;
    $S_{10} = 292378957513217070780644253/431938229218462612298455 \approx 0.67690$.

    (b) The sequence of partial sums is increasing because each term of the series is a positive number (i.e., $S_{n+1} = S_n + a_{n+1} > S_n$ because $a_{n+1} > 0$).

    Since $a_j < 3^{-j}$ for all $j \geq 0$, $S_n = \sum\limits_{j=0}^{n} a_j < \sum\limits_{j=0}^{n} 3^{-j} = \dfrac{1}{2}\left(3 - 3^{-n}\right) \leq \dfrac{3}{2}$ for all $n \geq 0$.

    Thus, each term in the sequence of partial sums is bounded above by $3/2$.

    (c) Since the sequence of partial sums is increasing and bounded above, it must converge. This implies that the infinite series converges.

5. (a) If $k \geq 1$, $k! = 1 \cdot \underbrace{2 \cdot 3 \cdots (k-1) \cdot k}_{k-1 \text{ terms}} \geq 1 \cdot \underbrace{2 \cdot 2 \cdots 2 \cdot 2}_{k-1 \text{ terms}} = 2^{k-1}$ and so $\dfrac{1}{k!} \leq \dfrac{1}{2^{k-1}}$.

    (b) $S_{n+1} = \sum\limits_{k=0}^{n+1} a_k = \sum\limits_{k=0}^{n} a_k + a_{n+1} = S_n + a_{n+1}$. Since $a_{n+1} > 0$, $S_{n+1} > S_n$.

    (c) Part (a) implies that

    $S_n = \sum\limits_{k=0}^{n} \dfrac{1}{k!} \leq 1 + \sum\limits_{k=1}^{n} \dfrac{1}{2^{k-1}} = 1 + \sum\limits_{k=0}^{n-1} \dfrac{1}{2^k} < 1 + \sum\limits_{k=0}^{\infty} \dfrac{1}{2^k} = 1 + 2 = 3$. Since the
    sequence of partial sums $\{S_n\}$ is increasing and bounded, the series converges.

6. (a) $a_1 = 1/1! = 1$, $a_2 = 1/2! = 1/2$, $a_5 = 1/5! = 1/120 \approx 0.0083333$,
    $a_{10} = 1/10! = 1/3628800 \approx 2.7557 \times 10^{-7}$.
    $S_1 = 2$; $S_2 = 2.5$; $S_5 = 163/60 \approx 2.71667$; $S_{10} = 9864101/3628800 \approx 2.71828$.

    (b) $a_{k+1} = \dfrac{1}{(k+1)!} = \dfrac{1}{(k+1)\,k!} < \dfrac{1}{k!} = a_k$.

    (c) Since $S_{n+1} = S_n + a_{n+1}$ and $a_{n+1} > 0$, $\{S_n\}$ is an increasing sequence.

(d) $R_n = \sum_{k=n+1}^{\infty} 1/k!$. Since each term in the sum is positive, $R_n > 0$.

(e) Since $R_n = e - S_n$ and S_n is a monotonically increasing sequence, $R_{n+1} < R_n$ for all $n \geq 0$.

(f) Any $n \geq 6$ will do.

(g) Any $n \geq 8$ will do.

(h) $R_{50} < R_8 < 10^{-5}$.

7. $\displaystyle\sum_{i=0}^{\infty} \frac{1}{(i+1)^4} = \sum_{j=1}^{\infty} \frac{1}{j^4} = \frac{\pi^4}{90}$.

8. $\displaystyle\sum_{k=3}^{\infty} \frac{1}{k^4} = \sum_{j=1}^{\infty} \frac{1}{j^4} - 1 - \frac{1}{2^4} = \frac{\pi^4}{90} - \frac{17}{16} \approx 0.019823$.

9. (a) When $r = 1$, $S_n = (n+1)a$, but the right-hand side of the expression is undefined (because of the zero in the denominator).

(b) $S_n - rS_n = (a + ar + ar^2 + \cdots + ar^n) - r(a + ar + ar^2 + \cdots + ar^n) = a - ar^{n+1} = a(1 - r^{n+1})$.

(c) $S_n - rS_n = (1 - r)S_n = a(1 - r^{n+1})$. The desired result is now obtained by dividing through by $1 - r$.

10. The sum is a geometric series with $a = 3$, $r = 2$, and $n = 10$. Thus,
$3 + 6 + 12 + \cdots + 3072 = 6141$.

11. $\displaystyle\frac{1}{16} + \frac{1}{32} + \frac{1}{64} + \frac{1}{128} + \cdots + \frac{1}{2^{i+4}} + \cdots = \frac{1}{2^4} \sum_{i=0}^{\infty} \left(\frac{1}{2}\right)^i = \frac{1}{2^4} \cdot 2 = \frac{1}{8}$.

12. $\displaystyle 2 - 5 + 9 + \frac{1}{3} + \frac{1}{9} + \frac{1}{27} + \frac{1}{81} + \cdots + \frac{1}{3^n} + \cdots = 6 + \frac{1}{3} \sum_{j=0}^{\infty} \left(\frac{1}{3}\right)^j = 6 + 1/2 = 6.5$.

13. $\displaystyle\sum_{n=0}^{\infty} e^{-n} = \frac{e}{e-1}$.

14. $\displaystyle\sum_{k=3}^{\infty} \left(\frac{e}{\pi}\right)^k = \sum_{k=0}^{\infty} \left(\frac{e}{\pi}\right)^k - \sum_{k=0}^{2} \left(\frac{e}{\pi}\right)^k = \frac{1}{1 - e/\pi} - 1 - \frac{e}{\pi} - \left(\frac{e}{\pi}\right)^2 = \frac{e^3}{\pi^2(\pi - e)} \approx 4.8076$.

15. $\displaystyle\sum_{m=2}^{\infty} (\arctan 1)^m = \sum_{m=2}^{\infty} (\pi/4)^m = \left(\frac{\pi}{4}\right)^2 \sum_{m=0}^{\infty} (\pi/4)^m = \frac{\pi^2}{16} \frac{1}{1 - \pi/4} = \frac{\pi^2}{16 - 4\pi}$.

16. $\displaystyle\sum_{i=10}^{\infty} \left(\frac{2}{3}\right)^i = \sum_{i=0}^{\infty} \left(\frac{2}{3}\right)^i - \sum_{i=0}^{9} \left(\frac{2}{3}\right)^i = \frac{1}{1 - (2/3)} - \frac{1 - (2/3)^{10}}{1 - (2/3)} = \frac{1024}{19683} \approx 0.052025$.

17. $\displaystyle\sum_{j=5}^{\infty}\left(-\frac{1}{2}\right)^{j} = \sum_{j=0}^{\infty}\left(-\frac{1}{2}\right)^{j} - \sum_{j=0}^{4}\left(-\frac{1}{2}\right)^{j}$

$\displaystyle = \frac{1}{1-(-1/2)} - \frac{1-(-1/2)^5}{1-(-1/2)} = -\frac{1}{48} \approx -0.020833.$

18. $\displaystyle\sum_{j=0}^{\infty}\frac{3^j+4^j}{5^j} = \sum_{j=0}^{\infty}\frac{3^j}{5^j} + \sum_{j=0}^{\infty}\frac{4^j}{5^j} = 5/2 + 5 = 15/2.$

19. Since $1/(2+\sin k) \geq 1/3$ for all $k \geq 1$, the series diverges by the nth-term test.

20. A series diverges if the sequence of its partial sums does not have a limit. Since $S_{2m} = 1$ and $S_{2m+1} = 0$ for $m = 0, 1, 2, 3, \ldots$, $\displaystyle\lim_{n\to\infty} S_n$ does not exist and, therefore, the series diverges.

21. $S_n = \arctan(n+1) - \arctan(0) = \arctan(n+1)$. Since $\displaystyle\lim_{n\to\infty} S_n = \frac{\pi}{2}$, the series converges to $\dfrac{\pi}{2}$.

22. $\displaystyle\sum_{j=1}^{\infty}\frac{j}{(j+1)!} = \sum_{j=1}^{\infty}\left(\frac{1}{j!} - \frac{1}{(j+1)!}\right)$ so $S_m = 1 - \dfrac{1}{(m+1)!}$. Since $\displaystyle\lim_{m\to\infty} S_m = 1$, the series converges to 1.

23. $S_n = 1 + \dfrac{1}{\sqrt{2}} - \dfrac{1}{\sqrt{n+1}} - \dfrac{1}{\sqrt{n+2}}$ when $n \geq 1$. Since $\displaystyle\lim_{n\to\infty} S_n = 1 + 1/\sqrt{2}$, the series converges to $1 + 1/\sqrt{2}$.

24. $S_n = \ln(n+1)$. Since $\displaystyle\lim_{n\to\infty} S_n = \infty$, the series diverges.

25. (a) $S_{100} = 5 - \dfrac{3}{100} = 4.97.$

    (b) $\displaystyle\sum_{k=1}^{\infty} a_k = \lim_{N\to\infty}\sum_{k=1}^{N} a_k = \lim_{N\to\infty} S_N = \lim_{N\to\infty}\left(5 - \frac{3}{N}\right) = 5.$

    (c) $\displaystyle\lim_{k\to\infty} a_k = 0$ since $\displaystyle\sum_{k=1}^{\infty} a_k$ converges (Theorem 6).

    (d) Since $a_1 = 2$, and $a_{n+1} = S_{n+1} - S_n = \left(5 - \dfrac{3}{n+1}\right) - \left(5 - \dfrac{3}{n}\right) = \dfrac{3}{n^2+n} > 0$ for all $n \geq 1$, $a_k > 0$ for all $k \geq 1$.

26. (a) H_n is the nth partial sum of the harmonic series. These partial sums form a monotonically increasing, divergent sequence.

    (b) $\displaystyle S_n = \sum_{k=0}^{n}\frac{1}{2k+1} \geq \sum_{k=0}^{n}\frac{1}{2k+2} = \frac{1}{2}\sum_{k=0}^{n}\frac{1}{k+1} = \frac{1}{2}\sum_{k=1}^{n+1}\frac{1}{k} = \frac{1}{2}H_{n+1} > \frac{1}{2}H_n.$

(c) $\displaystyle\sum_{k=0}^{\infty} \frac{1}{2k+1}$ diverges.

27. $\dfrac{1}{4} + \dfrac{1}{16} + \dfrac{1}{36} + \dfrac{1}{100} + \cdots = \dfrac{1}{4}\left(1 + \dfrac{1}{4} + \dfrac{1}{9} + \ldots\right) = \dfrac{\pi^2}{24}.$

28. $\displaystyle\sum_{k=0}^{\infty} \frac{1}{(2k+1)^2} = \sum_{k=1}^{\infty} \frac{1}{k^2} - \frac{1}{4}\sum_{k=1}^{\infty} \frac{1}{k^2} = \frac{\pi^2}{6} - \frac{\pi^2}{24} = \frac{\pi^2}{8}.$

29. $\displaystyle\sum_{m=1}^{\infty} \frac{(-1)^{m+1}}{m^2} = \sum_{m=1}^{\infty} \frac{1}{m^2} - \frac{1}{2}\sum_{m=1}^{\infty} \frac{1}{m^2} = \frac{\pi^2}{12}.$

Alternatively, $\displaystyle\sum_{m=1}^{\infty} \frac{(-1)^{m+1}}{m^2} = \sum_{k=0}^{\infty} \frac{1}{(2k+1)^2} - \sum_{j=1}^{\infty} \frac{1}{(2j)^2} = \frac{\pi^2}{8} - \frac{\pi^2}{24} = \frac{\pi^2}{12}.$

30. $\displaystyle\sum_{m=3}^{\infty} \frac{2^{m+4}}{5^m} = \frac{2^7}{5^3}\sum_{m=0}^{\infty} \left(\frac{2}{5}\right)^m = \frac{2^7}{5^3} \cdot \frac{5}{3} = \frac{128}{75}.$

31. The series converges for all values of x such that $-1 < x < 1$. $\displaystyle\sum_{k=0}^{\infty} x^k = \frac{1}{1-x}.$

32. The series converges when $-1 < x/5 < 1$. Thus, the series converges for all values of x such that $-5 < x < 5$. $\displaystyle\sum_{m=2}^{\infty} \left(\frac{x}{5}\right)^m = \sum_{m=0}^{\infty} \left(\frac{x}{5}\right)^m - \sum_{m=0}^{1} \left(\frac{x}{5}\right)^m = \frac{1}{1-(x/5)} - 1 - \frac{x}{5} = \frac{x^2}{5(5-x)}.$

33. Since $\displaystyle\sum_{j=5}^{\infty} x^{2j} = x^{10}\sum_{j=0}^{\infty} \left(x^2\right)^j$, the series converges when $-1 < x^2 < 1$. Thus, the series converges to $\dfrac{x^{10}}{1-x^2}$ for all values of x such that $-1 < x < 1$.

34. Since $\displaystyle\sum_{k=1}^{\infty} x^{-k} = \sum_{k=1}^{\infty} \left(x^{-1}\right)^k$, the series converges when $\left|x^{-1}\right| < 1$. Therefore, the series converges for all values of x such that $x < -1$ or $x > 1$. $\displaystyle\sum_{k=1}^{\infty} x^{-k} = \frac{1}{1-x^{-1}} - 1 = \frac{1}{x-1}.$

35. The series converges when $|1+x| < 1$. Thus, the series converges for all values of x such that $-2 < x < 0$. $\displaystyle\sum_{n=3}^{\infty} (1+x)^n = (1+x)^3\sum_{n=0}^{\infty} (1+x)^n = -\frac{(1+x)^3}{x}.$

36. The series converges when $\left|\dfrac{1}{1-x}\right| < 1$. Thus, the series converges for all values of x such that $x < 0$ or $x > 2$. $\displaystyle\sum_{j=4}^{\infty} \dfrac{1}{(1-x)^j} = \dfrac{x-1}{x(1-x)^4}$.

37. $S_{n+1} = \displaystyle\sum_{k=0}^{n}(1/3)^k \implies \lim_{n\to\infty} S_n = 3/2$.

38. $a_{n+1} = 4 - \displaystyle\sum_{k=1}^{n}(1/2)^k \implies \lim_{n\to\infty} a_n = 3$.

39. Diverges by the nth term test: $\displaystyle\lim_{n\to\infty}\dfrac{n+1}{2n+1} = \dfrac{1}{2} \neq 0$.

40. $\displaystyle\sum_{j=0}^{\infty}(\ln 2)^j = \dfrac{1}{1-\ln 2} \approx 3.2589$.

41. The partial sums of the series are
$$S_N = \sum_{n=2}^{N}\dfrac{2}{n^2-1} = \sum_{n=2}^{N}\left(\dfrac{1}{n-1} - \dfrac{1}{n+1}\right) = 1 + \dfrac{1}{2} - \dfrac{1}{N} - \dfrac{1}{N+1}.$$ Therefore, the series converges to $3/2$.

42. Diverges by the nth term test: $\displaystyle\lim_{n\to\infty}\left(1 + \dfrac{1}{n}\right)^n = e \neq 0$.

43. Diverges by the nth term test: $\displaystyle\lim_{n\to\infty}\sqrt[n]{\pi} = 1 \neq 0$.

44. Diverges. Each term of the series is a constant multiple $(1/\ln 10)$ of the corresponding term of the harmonic series since $\ln\left(10^k\right) = k\ln 10$.

45. Converges—by the comparison test:
$$\sum_{j=2}^{\infty}\dfrac{3^j}{4^{j+1}} = \dfrac{1}{4}\sum_{j=2}^{\infty}\left(\dfrac{3}{4}\right)^j = \dfrac{9}{64}\sum_{j=0}^{\infty}\left(\dfrac{3}{4}\right)^j = \dfrac{9}{64}\cdot\dfrac{1}{1-3/4} = \dfrac{9}{16}.$$
$$R_N = \sum_{j=N+1}^{\infty}\dfrac{3^j}{4^{j+1}} = \dfrac{3^{N+1}}{4^{N+2}}\sum_{j=0}^{\infty}\left(\dfrac{3}{4}\right)^j = \left(\dfrac{3}{4}\right)^{N+1} \leq 0.001 \text{ when}$$
$N \geq \dfrac{\ln 0.001}{\ln 3/4} - 1 \approx 23.012$. Thus, $n \geq 24$ implies that S_n approximates the sum of the series within 0.001.

46. Diverges. The given series is 2 minus the harmonic series:
$$1 - \dfrac{1}{2} - \dfrac{1}{3} - \dfrac{1}{4} - \dfrac{1}{5} - \cdots = 2 - \sum_{k=1}^{\infty}\dfrac{1}{k}.$$ Since the harmonic series diverges, so does this series.

47. Diverges—each term of this series is a constant multiple (1/100) of the corresponding term of the harmonic series.

48. Diverges by the nth term test.

49. The series $\displaystyle\sum_{k=0}^{\infty} \frac{3}{10}\left(-\frac{1}{2}\right)^k$ converges to 1/5.

50. Converges to 1. The partial sums are $S_{2n-1} = 1$ and $S_{2n} = 1 - 1/(n+1)$ $(n = 1, 2, 3, \ldots)$ so $\displaystyle\lim_{k\to\infty} S_k = 1$.

51. Diverges by the nth term test.

52. $\displaystyle\sum_{k=5}^{\infty} \frac{4}{7^{2k}} = \sum_{k=0}^{\infty} \frac{4}{49^k} - \sum_{k=0}^{4} \frac{4}{49^k} = \frac{4}{1-(1/49)} - \frac{4\left(1-(1/49)^5\right)}{1-(1/49)}$

$\displaystyle\qquad = \frac{49}{12} - \frac{23539604}{5764801} = \frac{1}{69177612}.$

53. After the first bounce, the ball rebounds to a height of $4 \cdot (2/3)$ feet. The ball then falls from this height and rebounds to $4 \cdot (2/3)^2$ feet, etc. Thus, the total distance that the ball travels is

$$4 + 2 \cdot 4 \cdot \frac{2}{3} + 2 \cdot 4 \cdot \left(\frac{2}{3}\right)^2 + 2 \cdot 4 \cdot \left(\frac{2}{3}\right)^3 + \cdots = 4 + 4 \cdot 2 \cdot \frac{2}{3} \cdot \frac{1}{1-(2/3)} = 20 \text{ feet.}$$

54. (a) $\displaystyle\lim_{k\to\infty} \frac{1}{\sqrt{k}} = 0.$

    (b) Since $1/\sqrt{n} \le 1/\sqrt{k}$ when $1 \le k \le n$, $\displaystyle\sum_{k=1}^{n} \frac{1}{\sqrt{k}} \ge \sum_{k=1}^{n} \frac{1}{\sqrt{n}} = \frac{n}{\sqrt{n}} = \sqrt{n}.$

    (c) It follows from part (b) that $\displaystyle\lim_{n\to\infty} S_n = \infty.$

55. Since $0 < \ln x \le \sqrt{x}$ for all $x \ge 2$, $\displaystyle\sum_{k=2}^{\infty} \frac{1}{\ln k} \ge \sum_{k=2}^{\infty} \frac{1}{\sqrt{k}}$. Now, the previous exercise implies that the series $\displaystyle\sum_{k=2}^{\infty} \frac{1}{\sqrt{k}}$ diverges. Therefore, the series $\displaystyle\sum_{k=2}^{\infty} \frac{1}{\ln k}$ also diverges.

56. (a) Yes. The terms of the sequence are increasing and bounded above, so Theorem 3 implies that the sequence has a limit.

    (b) Since the terms of the sequence are increasing, $a_k \ge a_1 > 0$ when $k \ge 1$. Therefore, $S_n = \sum_{k=1}^{n} a_k \ge \sum_{k=1}^{n} a_1 = na_1 > 0$ so $\displaystyle\lim_{n\to\infty} S_n = \infty$. This implies that the series diverges.

57. The series $\sum\limits_{k=1}^{\infty} a_k$ converges because its partial sums are bounded above (by 100) and increasing (the terms of the series are positive). Therefore, Theorem 6 implies that $\lim\limits_{k\to\infty} a_k = 0$.

58. (a) $\lim\limits_{n\to\infty} S_n = \ln 2$.

   (b) Yes. Since the sequence of partial sums has a limit, the series converges.

   (c) For any $n \geq 1$, $b_n = S_n - S_{n-1} = \ln\left(\dfrac{2n+3}{n+1}\right) - \ln\left(\dfrac{2n+1}{n}\right)$

$$= \ln\left(\dfrac{2n+3}{2n+1} \cdot \dfrac{n}{n+1}\right) = \ln\left(\dfrac{2n^2+3n}{2n^2+3n+1}\right) < 0.$$

59. (a) $\sum\limits_{k=1}^{\infty} a_k = \lim\limits_{n\to\infty} S_n = 3$ (by the Squeeze Theorem).

   (b) Since the series converges, $\lim\limits_{k\to\infty} a_k = 0$.

60. (a) $a_k = (-1)^{k+1}$.

   (b) By definition, the infinite series $\sum\limits_{k=1}^{\infty} a_k$ converges if the sequence of its partial sums has a limit—that is, the series converges if $\lim\limits_{n\to\infty} S_n$ exists. Since $a_k > 0$, S_n is an increasing sequence bounded above by 100 so Theorem 3 implies that this sequence has a limit and, therefore, that the infinite series converges.

   (c) $\sum\limits_{k=1}^{\infty} a_k = \sum\limits_{k=1}^{10^6-1} a_k + \sum\limits_{k=10^6}^{\infty} a_k.$ $\sum\limits_{k=1}^{10^6-1} a_k$ is a finite sum, so it is a real number. The infinite series $\sum\limits_{k=10^6}^{\infty} a_k$ converges because the sequence of its partial sums is increasing and bounded above. Thus, the original infinite series converges.

61. (a) Since $a_k \geq 0$, the partial sums of the series $\sum\limits_{k=1}^{\infty} a_k$ form a monotonically increasing sequence. Since the series diverges, the partial sums must increase without bound.

   (b) If $a_k = (-1)^k$, $S_n = -1$ when n is odd and $S_n = 0$ when n is even. Thus, $\lim\limits_{n\to\infty} S_n$ doesn't exist.

62. (a) If the interval $[1, n+1]$ is divided into n equal subintervals, each subinterval has a length of one and the left endpoint of the jth subinterval is $1/j$.

   (b) Since $1/x$ is a decreasing function on the interval $[1, \infty)$, $L_n > I_n$ for all $n \geq 1$. Now, $I_n = \ln(n+1) \implies \lim\limits_{n\to\infty} L_n = \infty$. Thus, since $H_n = L_n$, the sequence of partial sums of the harmonic series does not have a finite limit (i.e., the harmonic series diverges).

63. (a) $a_1 = \sum_{j=1}^{1} \frac{1}{1+j} = \frac{1}{2}$; $a_2 = \sum_{j=1}^{2} \frac{1}{2+j} = \frac{7}{12}$; $a_3 = \sum_{j=1}^{4} \frac{1}{4+j} = \frac{533}{840}$.

(b) $H_8 = H_{2^3} = 1 + \sum_{m=1}^{3} a_m = 1 + a_1 + a_2 + a_3 = \frac{761}{280}$.

(c) $a_k = \sum_{j=1}^{2^{k-1}} \frac{1}{2^{k-1} + j} \geq \sum_{j=1}^{2^{k-1}} \frac{1}{2^{k-1} + 2^{k-1}} = \frac{1}{2}$.

(d) $H_{2^n} = 1 + \sum_{k=1}^{n} a_k \geq 1 + \sum_{k=1}^{n} \frac{1}{2} = 1 + n/2 \implies \lim_{n \to \infty} H_{2^n} = \infty$.

64. (a) $S_{n+1} = S_n + \frac{1}{(n+1)^p} > S_n$ for all $n \geq 1$ since $\frac{1}{(n+1)^p} > 0$ for all $n \geq 1$.

(b) $S_{2m+1} = \sum_{k=1}^{2m+1} \frac{1}{k^p} = 1 + \sum_{k=2}^{2m+1} \frac{1}{k^p} = 1 + \sum_{k=1}^{m} \frac{1}{(2k)^p} + \sum_{k=1}^{m} \frac{1}{(2k+1)^p}$.

(c) $S_{2m+1} = 1 + \sum_{k=1}^{m} \frac{1}{(2k)^p} + \sum_{k=1}^{m} \frac{1}{(2k+1)^p}$

$\quad < 1 + \sum_{k=1}^{m} \frac{1}{(2k)^p} + \sum_{k=1}^{m} \frac{1}{(2k)^p} = 1 + 2\sum_{k=1}^{m} \frac{1}{(2k)^p}$

since $\frac{1}{(2k+1)^p} < \frac{1}{(2k)^p}$ for all $k \geq 1$ (since $p > 1$).

(d) $S_{2m+1} < 1 + 2\sum_{k=1}^{m} \frac{1}{(2k)^p} = 1 + 2^{1-p} \sum_{k=1}^{m} \frac{1}{k^p} = 1 + 2^{1-p} S_m < 1 + 2^{1-p} S_{2m+1}$. (The last inequality holds because the partial sums are strictly increasing.)

(e) By part (d), $S_{2m+1} < 1 + 2^{1-p} S_{2m+1}$. Therefore, $(1 - 2^{1-p}) S_{2m+1} < 1$ so $S_{2m+1} < \frac{1}{1 - 2^{1-p}}$ for every $m \geq 1$. Finally, since any $n \geq 1$ can be written as either $n = 2m$ or $n = 2m + 1$ where $m \geq 1$ is an integer, and since $S_{2m} < S_{2m+1}$ for all $m \geq 1$, it follows that sequence of partial sums is bounded above.

Since the sequence of partial sums of the series is increasing and bounded above, this sequence has a finite limit. By definition, this implies that the series converges.

§11.3   Testing for Convergence; Estimating Limits

1. (a) When $k \geq 0$, $k + 2^k \geq 2^k \implies a_k \leq 1/2^k = 2^{-k}$. Since $\displaystyle\sum_{k=0}^{\infty} 2^{-k}$ converges, the

    comparison test implies that $\displaystyle\sum_{k=0}^{\infty} a_k$ converges.

    (b) $R_{10} = \displaystyle\sum_{k=11}^{\infty} a_k < \sum_{k=11}^{\infty} 2^{-k} = 2^{-11} \sum_{k=0}^{\infty} 2^{-k} = 2^{-10}$. Furthermore, since $a_k > 0$ for all
    $k \geq 11$, $R_{10} \geq 0$.

    (c) Since $R_{10} < 2^{-10} \approx 0.00097656$, S_n has the desired accuracy if $n \geq 10$.
    $$S_{10} = \sum_{k=0}^{10} a_k = \frac{127807216183}{75344540040} \approx 1.6963.$$

    (d) No. Since the terms of the series are all positive, the estimate in part (c) *underestimates* the limit.

2. (a) The series converges by the comparison test: $\dfrac{1}{2 + 3^j} < 3^{-j}$ for all $j \geq 0$ and $\displaystyle\sum_{j=0}^{\infty} 3^{-j}$
    converges.

    (b) $R_n = \displaystyle\sum_{j=n+1}^{\infty} \frac{1}{2 + 3^j} < \sum_{j=n+1}^{\infty} 3^{-j} = \frac{1}{2 \cdot 3^n}$. Since $R_n < 0.01$ for all $n \geq 4$, the partial
    sum
    $S_4 = 266401/397155 \approx 0.67077$ approximates the limit with the desired accuracy.

    (c) No. Since the terms of the series are all positive, the estimate in part (c) *underestimates* the limit.

3. $\displaystyle\sum_{k=2}^{n} a_k < \int_{1}^{n} a(x)\, dx < \sum_{k=1}^{n-1} a_k$.   [HINT: Draw pictures like those on p. 569.]

4. $\displaystyle\int_{n+1}^{\infty} a(x)\, dx < \sum_{k=n+1}^{\infty} a_k < \int_{n}^{\infty} a(x)\, dx.$

5. Draw a picture illustrating the left sum approximation L_n to the integral $\displaystyle\int_{1}^{n+1} a(x)\, dx$.
    Since the integrand is a decreasing function, L_n overestimates the value of the integral.

6. The right Riemann sum approximation R_{n-1} of the integral $\displaystyle\int_{1}^{n} a(x)\, dx$ is $\displaystyle\sum_{k=2}^{n} a_k$. The
    approximation underestimates the value of the integral since the integrand is a decreasing
    function.

7. Draw a picture that illustrates a right Riemann sum approximation to the integral
$\int_n^\infty a(x)\,dx$. The right sum is $\sum_{k=n+1}^\infty a_k$; it underestimates $\int_n^\infty a(x)\,dx$ since the integrand is
a decreasing function. For the same reason, $a_{n+1} + \int_{n+1}^\infty a(x)\,dx \le \int_n^\infty a(x)\,dx$. Finally,
the first inequality established in this solution implies that $\sum_{k=n+2}^\infty a_k \le \int_{n+1}^\infty a(x)\,dx$. Adding
a_{n+1} to both sides of this inequality establishes that
$$\sum_{k=n+1}^\infty a_k \le a_{n+1} + \int_{n+1}^\infty a(x)\,dx.$$

8. The sum on the left side of the inequality is a right sum approximation to the integral on the
right side of the inequality. Since the integrand is decreasing, the right sum approximation
underestimates the value of the integral.

9. $\int_1^\infty \dfrac{dx}{x^3} \le \sum_{k=1}^\infty \dfrac{1}{k^3} \le 1 + \int_1^\infty \dfrac{dx}{x^3} \implies \dfrac{1}{2} \le \sum_{k=1}^\infty \dfrac{1}{k^3} \le \dfrac{3}{2}.$

10. $\int_1^\infty \dfrac{dx}{x^{3/2}} = -\dfrac{2}{\sqrt{x}}\bigg|_1^\infty = 2 \implies \sum_{k=1}^\infty \dfrac{1}{k\sqrt{k}}$ converges and $2 \le \sum_{k=1}^\infty \dfrac{1}{k\sqrt{k}} \le 3.$

11. $\int_1^\infty x e^{-x}\,dx = -e^{-x}(x+1)\bigg|_1^\infty = 2e^{-1} \implies \sum_{j=1}^\infty j e^{-j}$ converges and
$2e^{-1} \le \sum_{j=1}^\infty j e^{-j} \le 3e^{-1}.$

12. $\int_1^\infty \dfrac{dx}{x^2 + 1} = \arctan x \bigg|_1^\infty = \pi/4 \implies \sum_{k=0}^\infty \dfrac{1}{k^2 + 1}$ converges and
$1 + \dfrac{\pi}{4} \le 1 + \sum_{k=1}^\infty \dfrac{1}{k^2 + 1} \le 1 + \dfrac{1}{2} + \dfrac{\pi}{4}.$

13. (a) No. The function $a(x) = e^{\sin x}/x^2$ is *not* decreasing on the interval $[1, \infty)$ so it does
not satisfy the hypotheses of the theorem.

(b) For all $k \ge 1, 0 < \dfrac{e^{\sin k}}{k^2} \le \dfrac{e}{k^2}$. Therefore, since $\sum_{k=1}^\infty \dfrac{e}{k^2} = e \sum_{k=1}^\infty \dfrac{1}{k^2}$ converges, the
comparison test implies that $\sum_{k=1}^\infty \dfrac{e^{\sin k}}{k^2}$ converges.

14. If $x \geq 2, 0 < \ln x < \sqrt{x} \implies (\ln x)^2 < x \implies \sum_{n=2}^{\infty} \frac{1}{(\ln n)^2} > \sum_{n=2}^{\infty} \frac{1}{n}$. Since the series on the right diverges, so must the one on the left.

15. Since $\int_{2}^{\infty} \frac{dx}{(\ln x)^2} \geq \int_{2}^{\infty} \frac{dx}{x \ln x} = \infty$, the series $\sum_{n=2}^{\infty} \frac{1}{(\ln n)^2}$ diverges by the integral test.

16. $\sum_{k=3}^{\infty} \frac{1}{(\ln k)^k} < \sum_{k=3}^{\infty} \frac{1}{(\ln 3)^k} = \frac{1}{(\ln 3)^3 (1 - 1/\ln 3)} \approx 8.40195$. Thus, the series converges by the comparison test.

17. $S_1 = \frac{1}{2} < \sum_{n=1}^{\infty} \frac{1}{n^2 + \sqrt{n}} < \sum_{k=1}^{\infty} \frac{1}{n^2} \leq 2.$

18. $1 = \frac{1}{0 + e^0} < \sum_{j=0}^{\infty} \frac{1}{j + e^j} < \sum_{j=0}^{\infty} \frac{1}{e^j} = \frac{1}{1 - e^{-1}}.$ (The series on the right is a convergent geometric series.)

19. $\frac{1}{\sqrt{2}} = \frac{1}{\sqrt{1 + 1^2}} < \sum_{m=1}^{\infty} \frac{1}{m\sqrt{1 + m^2}} < \sum_{m=1}^{\infty} \frac{1}{m^2} \leq 2.$

20. $\frac{1}{4} = \frac{1}{\left(1^2 + 1\right)^2} < \sum_{k=1}^{\infty} \frac{k}{\left(k^2 + 1\right)^2} < \sum_{k=1}^{\infty} \frac{1}{k^3} \leq 1 + \int_{1}^{\infty} x^{-3} \, dx = \frac{3}{2}.$

21. Let $a_j = \frac{j^2}{j!}$. Since
$$\lim_{j \to \infty} \frac{a_{j+1}}{a_j} = \lim_{j \to \infty} \frac{\frac{(j+1)^2}{(j+1)!}}{\frac{j^2}{j!}} = \lim_{j \to \infty} \frac{(j+1)^2}{j^2} \cdot \frac{j!}{(j+1)!} = \lim_{j \to \infty} \frac{j+1}{j^2} = 0 < 1, \sum_{j=0}^{\infty} a_j$$
converges.

22. $\lim_{k \to \infty} \frac{\frac{2^{k+1}}{(k+1)!}}{\frac{2^k}{k!}} = \lim_{k \to \infty} \frac{2}{k+1} = 0 \implies \sum_{k=1}^{\infty} \frac{2^k}{k!}$ converges.

23. $\lim_{n \to \infty} \frac{\frac{(n+1)^2}{2^{n+1}}}{\frac{n^2}{2^n}} = \lim_{n \to \infty} \left(\frac{n+1}{n}\right)^2 \cdot \frac{1}{2} = \frac{1}{2} \implies \sum_{n=1}^{\infty} \frac{n^2}{2^n}$ converges.

24. $\lim_{m \to \infty} \frac{\frac{(m+1)!}{(2m+2)!}}{\frac{m!}{(2m)!}} = \lim_{m \to \infty} \frac{m+1}{(2m+2)(2m+1)} = 0 \implies \sum_{m=1}^{\infty} \frac{m!}{(2m)!}$ converges.

25. (a) $\ln x / x$ is decreasing on $[3, \infty)$ and $\int_3^\infty \dfrac{\ln x}{x} \, dx$ diverges, so the integral test implies that $\displaystyle\sum_{k=3}^\infty \dfrac{\ln k}{k}$ diverges. Therefore, since $\displaystyle\sum_{k=1}^\infty \dfrac{\ln k}{k} = \dfrac{1}{2} \ln 2 + \sum_{k=3}^\infty \dfrac{\ln k}{k}$, $\displaystyle\sum_{k=1}^\infty \dfrac{\ln k}{k}$ diverges.

    (b) Since $1 - 1/k \le \ln k$ for all $k \ge 1$, $1/k - 1/k^2 \le (\ln k)/k$ for all $k \ge 1$. Therefore, since $\displaystyle\sum_{k=1}^\infty \left(\dfrac{1}{k} - \dfrac{1}{k^2} \right)$ diverges, $\displaystyle\sum_{k=1}^\infty \dfrac{\ln k}{k}$ diverges.

    (c) No, because $\displaystyle\lim_{k \to \infty} \dfrac{a_{k+1}}{a_k} = 1$.

26. (a) $\displaystyle\lim_{n \to \infty} \dfrac{a_{n+1}}{a_n}$ does not exist. For $m = 1, 2, 3, \ldots$, $\dfrac{a_{2m}}{a_{2m-1}} = \left(\dfrac{2}{3} \right)^m$ and $\dfrac{a_{2m+1}}{a_{2m}} = \dfrac{1}{2} \left(\dfrac{3}{2} \right)^m$.

    (b) Because the limit in part (a) does not exist, the ratio test says nothing about the convergence of the series $\displaystyle\sum_{k=1}^\infty a_k$.

    (c) $\displaystyle\sum_{k=1}^\infty a_k = \sum_{k=1}^\infty 2^{-k} + \sum_{k=1}^\infty 3^{-k} = 1 + \dfrac{1}{2} = \dfrac{3}{2}$.

27. (a) The ratio $\dfrac{a_{k+1}}{a_k}$ is 1 when k is odd and $1/2$ when k is even. Thus, $\displaystyle\lim_{k \to \infty} \dfrac{a_{k+1}}{a_k}$ doesn't exist.

    (b) It converges. Let S_n denote the partial sum of the first n terms of the series. When $n = 2m$, $S_n = 2 - \left(\dfrac{1}{2} \right)^{m-1}$; when $n = 2m + 1$, $S_n = 2 - 3 \left(\dfrac{1}{2} \right)^{m+1}$. It follows that $\displaystyle\lim_{n \to \infty} S_n = 2$.

28. The harmonic series $\displaystyle\sum_{k=1}^\infty \dfrac{1}{k}$ has the specified properties.

29. Let $a_n = n^{-n}$. Then $\dfrac{a_{n+1}}{a_n} = \dfrac{(n+1)^{-(n+1)}}{n^{-n}} = \dfrac{n^n}{(n+1)^{(n+1)}} = \left(\dfrac{n}{n+1} \right)^n \dfrac{1}{n+1}$. Since $\displaystyle\lim_{n \to \infty} \left(\dfrac{n}{n+1} \right)^n = \dfrac{1}{e}$ and $\displaystyle\lim_{n \to \infty} \dfrac{1}{n+1} = 0$, $\displaystyle\lim_{n \to \infty} \dfrac{a_{n+1}}{a_n} = 0$. Thus, the ratio test implies that the series $\displaystyle\sum_{n=1}^\infty n^{-n}$ converges.

30. Let $a_n = \dfrac{n^n}{n!}$. Then $\dfrac{a_{n+1}}{a_n} = \dfrac{(n+1)^{n+1}}{n^n} \dfrac{n!}{(n+1)!} = \left(\dfrac{n+1}{n} \right)^n$ and $\displaystyle\lim_{n \to \infty} \dfrac{a_{n+1}}{a_n} = e > 1$. Therefore, the series $\displaystyle\sum_{n=1}^\infty \dfrac{n^n}{n!}$ diverges.

31. Converges—by the comparison test: $\displaystyle\sum_{k=1}^{\infty} \frac{1}{k^2 + 3} \leq \sum_{k=1}^{\infty} \frac{1}{k^2}$. (The series on the right side of the inequality is a convergent p-series.)

    Since $\displaystyle R_N = \sum_{k=N+1}^{\infty} \frac{1}{k^2 + 3} \leq \sum_{k=N+1}^{\infty} \frac{1}{k^2} \leq \int_N^{\infty} \frac{dx}{x^2} = \frac{1}{N} \leq 0.001$ when $N \geq 1000$,

    $n \geq 1000$ implies that S_n approximates the sum of the series within 0.001.

32. Diverges—by the comparison test: $\displaystyle\sum_{m=1}^{\infty} \frac{\arctan m}{m} \geq \frac{\pi}{4} \sum_{m=1}^{\infty} \frac{1}{m}$. (The series on the right hand side of the inequality is the harmonic series—a series known to diverge.)

    $\displaystyle S_n = \sum_{m=1}^{n} \frac{\arctan m}{m} \geq \frac{\pi}{4} \sum_{m=1}^{n} \frac{1}{m} \geq \frac{\pi}{4} \int_1^{n+1} \frac{dx}{x} = \frac{\pi}{4} \ln(n+1) \geq 1000$ if

    $n \geq e^{4000/\pi} - 1 \approx 9.2 \times 10^{552}$.

33. Diverges—by the nth term test: $\displaystyle\lim_{k \to \infty} \frac{1}{2 + \cos k}$ doesn't exist.

    $\displaystyle S_n = \sum_{k=0}^{n} \frac{1}{2 + \cos k} \geq \sum_{k=0}^{n} \frac{1}{3} = \frac{n+1}{3}$, so $S_n \geq 1000$ when $n \geq 2999$.

34. Converges—by the integral test:

    $$\sum_{m=2}^{\infty} \frac{\ln m}{m^3} \leq \frac{\ln 2}{8} + \int_2^{\infty} \frac{\ln x}{x^3} \, dx = \frac{\ln 2}{8} + \frac{1 + 2\ln 2}{16}.$$

    $$0 < R_n = \sum_{m=n+1}^{\infty} \frac{\ln m}{m^3} \leq \int_n^{\infty} \frac{\ln x}{x^3} \, dx \leq \int_n^{\infty} \frac{x}{x^3} \, dx = \frac{1}{n} \leq 0.001 \text{ when } n \geq 1000.$$

35. The comparison $\displaystyle\sum_{k=1}^{\infty} \frac{k}{k^6 + 17} < \sum_{k=1}^{\infty} \frac{k}{k^6} = \sum_{k=1}^{\infty} \frac{1}{k^5}$ shows that the original series converges.

    Since $\displaystyle R_n \leq \int_n^{\infty} \frac{x}{x^6 + 17} \, dx < \int_n^{\infty} \frac{dx}{x^5} = \frac{1}{4n^4} < 0.001$ when $n \geq 4$, the estimate

    $\displaystyle\sum_{k=1}^{4} \frac{k}{k^6 + 17} \approx 0.08524$ is guaranteed to be in error by no more than 0.001.

36. Since the improper integral $\displaystyle\int_2^{\infty} \frac{1}{x(\ln x)^5} \, dx = \int_{\ln 2}^{\infty} \frac{1}{u^5} = \frac{1}{4(\ln 2)^4}$ converges, the

    corresponding series converges. Furthermore, since $\displaystyle\int_n^{\infty} \frac{dx}{x(\ln x)^5} \, dx < 0.001$ when $n \geq 54$,

    the approximation $\displaystyle\sum_{k=2}^{54} \frac{1}{k(\ln k)^5} \approx 3.4288$ has the desired accuracy.

37. (a) Let $M = L + 1$ where $L = \sum_{k=1}^{\infty} b_k = \lim_{n \to \infty} T_n$. Since $0 \le a_k \le b_k$ for all $k \ge 1$, $S_n \le T_n$ for all $n \ge 1$. Furthermore, since $0 \le b_k$ for all $k \ge 1$, $\{T_n\}$ is an increasing sequence whose limit is $L < M$. It follows that $S_n \le T_n < M$ for all $n \ge 1$.

(b) S_n is an increasing sequence because $0 \le a_k$ for all $k \ge 1$: $S_{n+1} = S_n + a_{n+1} > S_n$.

(c) Together, parts (a) and (b) imply that $\{S_n\}$ is an increasing sequence that is bounded above. Since $\{S_n\}$ is a sequence of partial sums of the infinite series $\sum_{k=1}^{\infty} a_k$, convergence of the sequence implies convergence of the series.

38. (a) Divergence of the series implies that the sequence of its partial sums diverges. Since $0 \le a_k$ for all $k \ge 1$, $S_n \ge 0$ for all $n \ge 1$ and $\{S_n\}$ is an increasing sequence.

(b) Since $S_n \le T_n$ for all $n \ge 1$, $\lim_{n \to \infty} S_n = \infty \implies \lim_{n \to \infty} T_n = \infty$.

39. (a) $S_{n+1} = \sum_{k=1}^{n+1} a_k = S_n + a_{n+1} \ge S_n$ since $a_k = a(k) \ge 0$ for all integers $k \ge 1$.

(b) $\int_{1}^{n} a(x)\,dx \le \int_{1}^{\infty} a(x)\,dx$ because $a(x) \ge 0$ for all $x \ge 1$.

(c) Part (a) implies that $\{S_n\}$ is an increasing sequence. Since $S_n \le a_1 + \int_{1}^{n} a(x)\,dx$, part (b) implies that the sequence of partial sums is bounded above. Thus, the sequence of partial sums converges to a limit.

40. (a) The assumption that $a(x)$ is decreasing is necessary to ensure that the desired geometric relationships hold (e.g., that $\int_{1}^{\infty} a(x)\,dx \le \sum_{k=1}^{\infty} a_k$).

(b) Given the new assumption, the inequality in the second conclusion must be replaced by $\int_{10}^{\infty} a(x)\,dx \le \sum_{k=10}^{\infty} a_k \le a_{10} + \int_{10}^{\infty} a(x)\,dx$, and the condition $n \ge 10$ must be placed on the inequality in the last conclusion. No other changes to the theorem are necessary.

41. **Yes.** Let the series in question be $\sum_{k=0}^{\infty} a_k$ (i.e., $a_0 = 1$, $a_1 = 1/3$, $a_2 = 1/15$, $a_3 = 1/105$, etc.). Then, $a_k \le 3^{-k}$. Since $\sum_{k=0}^{\infty} 3^{-k}$ is a convergent geometric series, the series $\sum_{k=0}^{\infty} a_k$ converges by the comparison test.

42. Yes — use the comparison test with the geometric series $\sum_{k=1}^{\infty} 2^{-k}$ or the series $\frac{1}{2} \sum_{k=0}^{\infty} \frac{1}{k!}$.

43. $0 < \sum_{k=3}^{\infty} \dfrac{1}{(\ln k)^{\ln k}} = \sum_{k=3}^{1619} \dfrac{1}{(\ln k)^{\ln k}} + \sum_{k=1620}^{\infty} \dfrac{1}{(\ln k)^{\ln k}}$

$< \sum_{k=3}^{1619} \dfrac{1}{(\ln k)^{\ln k}} + \sum_{k=1620}^{\infty} \dfrac{1}{e^{2\ln k}}$

$< \sum_{k=3}^{1619} \dfrac{1}{(\ln k)^{\ln k}} + \sum_{k=1620}^{\infty} \dfrac{1}{k^2}.$

Since the last series on the right converges (it is a p-series), the desired result follows by the comparison test.

44. Since $a_n \geq 0$ for all $n \geq 1$, $0 \leq |\sin(a_n)| \leq |a_n| = a_n$ and $0 \leq \sum_{n=1}^{\infty} |\sin(a_n)| \leq \sum_{n=1}^{\infty} a_n$.

Therefore, $\sum_{n=1}^{\infty} |\sin(a_n)|$ converges (by the comparison test).

45. Converges. $\displaystyle\int_1^{\infty} \dfrac{\arctan x}{1+x^2}\, dx \leq \sum_{n=1}^{\infty} \dfrac{\arctan n}{1+n^2} \leq \dfrac{\pi}{8} + \int_1^{\infty} \dfrac{\arctan x}{1+x^2}\, dx.$ Now,

$\displaystyle\int_1^{\infty} \dfrac{\arctan x}{1+x^2}\, dx = \int_{\pi/4}^{\pi/2} u\, du = \dfrac{3\pi^2}{32}.$ Therefore, $\dfrac{3\pi^2}{32} \leq \sum_{n=1}^{\infty} \dfrac{\arctan n}{1+n^2} \leq \dfrac{\pi}{8} + \dfrac{3\pi^2}{32}.$

[NOTE: : Since the terms of the series are positive, any partial sum could also be used to provide a lower bound on the limit of the series. Thus, $a_1 = \pi/8$ could also be used as a lower bound.]

46. Converges—by the comparison test: $\dfrac{1}{4} < \sum_{m=1}^{\infty} \dfrac{m^3}{m^5+3} < \sum_{m=1}^{\infty} \dfrac{m^3}{m^5} = \sum_{m=1}^{\infty} \dfrac{1}{m^2} \leq 2.$

[NOTE: : Since the terms of the series are positive, any partial sum could be used to provide a lower bound on the limit of the series.]

47. Diverges—by the integral test: $\displaystyle\int_1^{\infty} \dfrac{dx}{100+5x} = \dfrac{1}{5} \int_{105}^{\infty} \dfrac{du}{u} = \infty.$

48. Diverges—by the integral test: $\displaystyle\int_2^{\infty} \dfrac{dx}{x \ln x} = \int_{\ln 2}^{\infty} \dfrac{du}{u} = \infty.$

49. Converges—by the comparison test: $\dfrac{1}{3} < \sum_{n=1}^{\infty} \dfrac{1}{n \cdot 3^n} < \sum_{n=1}^{\infty} \dfrac{1}{3^n} = \dfrac{1}{2}.$

[NOTE: : Since the terms of the series are positive, any partial sum could be used to provide a lower bound on the limit of the series.]

50. Diverges—by the comparison test: $\displaystyle\sum_{n=2}^{\infty} \dfrac{1}{\sqrt[3]{n^2-1}} > \sum_{n=2}^{\infty} \dfrac{1}{n^{2/3}}.$

51. Converges—by the comparison test:
$$\sum_{j=0}^{\infty} \frac{j!}{(j+2)!} = \sum_{j=0}^{\infty} \frac{1}{(j+2)(j+1)}$$

$$= \frac{1}{2} + \sum_{j=1}^{\infty} \frac{1}{(j+2)(j+1)}$$

$$< \frac{1}{2} + \sum_{j=1}^{\infty} \frac{1}{j^2}$$

$$\leq \frac{1}{2} + 1 + \int_{1}^{\infty} \frac{dx}{x^2} = \frac{5}{2}.$$

Thus, $\dfrac{1}{2} < \displaystyle\sum_{j=0}^{\infty} \frac{j!}{(j+2)!} \leq \frac{5}{2}.$

[NOTE: : Since the terms of the series are positive, any partial sum could be used to provide a lower bound on the limit of the series.]

52. Converges—by the ratio test:
$$\lim_{n\to\infty} \frac{\frac{(n+1)!}{(2n+2)!}}{\frac{n!}{(2n)!}} = \lim_{n\to\infty} \frac{(n+1)!}{n!} \cdot \frac{(2n)!}{(2n+2)!} = \lim_{n\to\infty} \frac{1}{2(2n+1)} = 0.$$

Since $\dfrac{a_{n+1}}{a_n} = \dfrac{1}{2(2n+1)} \leq \dfrac{1}{2}$ for all $n \geq 0$, $\displaystyle\sum_{n=0}^{\infty} a_n \leq a_0 \sum_{n=0}^{\infty} \frac{1}{2^n} = 2a_0 = 2.$ Thus,

$$1 \leq \sum_{n=0}^{\infty} \frac{n!}{(2n)!} \leq 2.$$

Alternatively, since $\dfrac{n!}{(2n)!} \leq \dfrac{1}{(n+1)!}$ when $n \geq 0$, the series can also be shown to converge

using the comparison test: $\displaystyle\sum_{n=0}^{\infty} \frac{n!}{(2n)!} < \sum_{n=0}^{\infty} \frac{1}{(n+1)!} = e - 1.$

[NOTE: : Since the terms of the series are positive, any partial sum could be used to provide a lower bound on the limit of the series.]

53. (a) The exercise statement implies that $a_2/a_1 \leq r$ and $a_1 > 0$. Therefore, $a_2 \leq a_1 r$.

(b) $\dfrac{a_{k+1}}{a_k} \leq r < 1$ for all $k \geq 1$ implies that $a_{k+1} \leq a_k r < a_k$. From this it follows that $a_2 \leq a_1 r$ and $a_3 \leq a_2 r$. Multiplying both sides of the inequality $a_2 \leq a_1 r$ by r produces $a_2 r \leq a_1 r^2$. Therefore, $a_3 \leq a_2 r \leq a_1 r^2$.

[NOTE: Multiplying both sides of the inequality $a_3 \leq a_1 r^2$ by r leads to $a_3 r \leq a_1 r^3$. Since $a_4 \leq a_3 r$, we may conclude that $a_4 \leq a_1 r^3$. Proceeding in a similar fashion, one finds that $a_{k+1} \leq a_1 r^k$.]

(c) $\displaystyle\sum_{k=1}^{\infty} a_k \leq \sum_{k=0}^{\infty} a_1 r^k = \frac{a_1}{1-r}.$ This implies that the given series converges.

(d) $a_{n+k} \leq a_{n+1} r^{k-1}$ for all $k \geq 1$ so $R_n = \sum\limits_{k=n+1}^{\infty} a_k \leq a_{n+1} \sum\limits_{k=0}^{\infty} r^k = \dfrac{a_{n+1}}{1-r}$.

54. Let $a_n = n^2/2^n$. Since $a_{n+1}/a_n = \dfrac{(n+1)^2}{2n^2} \leq 8/9$ when $n \geq 3$,

$$R_N = \sum_{n=N+1}^{\infty} a_n \leq a_{N+1} + \tfrac{8}{9} a_{N+1} + \left(\tfrac{8}{9}\right)^2 a_{N+1} + \cdots = a_{N+1} \sum_{n=0}^{\infty} \left(\frac{8}{9}\right)^n = 9 a_{N+1}.$$

Thus, $R_N < 0.0005$ when $N \geq 23$.

55. Let $a_n = (n!)^2/(2n)!$. Since $a_{n+1}/a_n = \dfrac{n+1}{2(2n+1)} \leq \tfrac{1}{3}$ when $n \geq 1$,

$R_N = \sum\limits_{n=N+1}^{\infty} a_n \leq a_{N+1} \sum\limits_{n=0}^{\infty} \left(\dfrac{1}{3}\right)^n = \dfrac{3}{2} a_{N+1}$. Thus, $R_N < 0.0005$ when $N \geq 6$.

§11.4 Absolute Convergence; Alternating Series

1. The series converges conditionally. (After the fifth term, the series has the same terms as the alternating harmonic series.)

2. $S_{15} = 1 + 2 + 3 + 4 + 5 + \sum_{k=6}^{15} \frac{(-1)^{k+1}}{k} = 15 - \frac{20887}{360360} \approx 14.942.$ S_{15} *overestimates* S because the last term included in the alternating series was positive.

3. $14.902 < S < 14.902 + \frac{1}{61} \approx 14.918.$

4. $S = 15 + \left(\ln 2 - \sum_{k=1}^{5} \frac{(-1)^{k+1}}{k} \right) = 15 + \ln 2 - \frac{47}{60} \approx 14.910.$

5. (a) $S_{50} \approx 0.23794.$

   (b) $|R_{50}| = \left| \sum_{n=51}^{\infty} \frac{\sin n}{n^3 + n^2 + n + 1 + \cos n} \right| \le \sum_{n=51}^{\infty} \left| \frac{\sin n}{n^3 + n^2 + n + 1 + \cos n} \right|$

   $\le \sum_{n=51}^{\infty} \frac{1}{n^3 + n^2 + n + 1 + \cos n} \le \sum_{n=51}^{\infty} \frac{1}{n^3} \le \int_{50}^{\infty} \frac{dx}{x^3}.$

   (The last step follows from the integral test.)

   (c) By part (b), $|R_{50}| \le \frac{1}{5000} = 0.0002.$ Therefore, using the result in part (a),
   $S_{50} - 0.0002 \approx 0.23774 < S < S_{50} + 0.0002 \approx 0.23814.$

6. $0 < \sum_{k=1}^{\infty} \frac{|a_k|}{k} < \sum_{k=1}^{\infty} |a_k|,$ so $\sum_{k=1}^{\infty} \frac{a_k}{k}$ converges absolutely. This implies (by Theorem 10) that this series converges.

7. No — An example is $a_k = 1/k.$ Then, $\sum_{k=1}^{\infty} a_k/k = \sum_{k=1}^{\infty} 1/k^2$ (a convergent series), but

   $\sum_{k=1}^{\infty} a_k = \sum_{k=1}^{\infty} 1/k$ (the divergent harmonic series).

8. Let $a_k = k/(k^2 - 1).$ Since $\lim_{k \to \infty} a_k = 0$ and $a_{k+1} < a_k$ for all $k \ge 2,$ Theorem 11 implies

   that $\sum_{k=2}^{\infty} (-1)^k a_k$ converges. However, $\sum_{k=2}^{\infty} a_k$ diverges ($a_k \ge 1/k$), so the alternating series converges conditionally.

9. The series converges absolutely by the alternating series test. Since

   $c_{n+1} = (n+1)^{-4} < 0.005$ when $n = 3,$ $\left| S - \sum_{k=1}^{N} (-1)^k / k^4 \right| < 0.005$ when $N \ge 3.$ Using

   $N = 3,$ $S \approx -\frac{1231}{1296} \approx -0.94985.$

10. The series converges absolutely by the comparison test using $b_k = 2^{-k}$. $\left| S - \sum_{k=1}^{N} a_k \right| \le 0.005$

    when $N \ge 7$ since $1/(8^2 + 2^8) < 0.005$. Using $N = 7$, $S \approx -\dfrac{11393057}{45736800} \approx -0.24910$.

11. The series converges absolutely by the ratio test. $\left| S - \sum_{k=0}^{N} a_k \right| \le 0.005$ when $N \ge 2$ since

    $3^3/9! < 0.005$. Using $N = 2$, $S \approx -\dfrac{13}{8} = -1.625$.

12. The series converges absolutely by the ratio test. $\left| S - \sum_{k=5}^{N} a_k \right| \le 0.005$ when $N \ge 13$ since

    $14^{10}/10^{14} < 0.005$. Using $N = 13$, $S \approx -\dfrac{573982077919709}{10000000000000} \approx -57.398$.

13. **No.** Since $a_k \ge 0$ for all $k \ge 1$, $|a_k| = a_k$ for all $k \ge 1$.

14. No. Because $\lim\limits_{n \to \infty} (-1)^n \dfrac{n}{2n - 1}$ does not exist, the series cannot converge (by the n-th term test).

15. When $p \le 0$, $\lim\limits_{k \to \infty} \dfrac{\ln k}{k^p} = \infty$, so the series diverges by the nth term test (Theorem 6).

    When $p > 0$, the function $f(x) = \dfrac{\ln x}{x^p}$ is continuous, positive, and decreasing on $(e^{1/p}, \infty)$. Therefore, the integral test (Theorem 8) implies that the series converges if and only if the improper integral $\displaystyle\int_{e^{1/p}}^{\infty} f(x)\, dx$ converges.

    When $p \ne 1$, $\displaystyle\int \dfrac{\ln x}{x^p}\, dx = \dfrac{(1 - p) \ln x - 1}{(1 - p)^2 x^{p-1}}$ and, $\displaystyle\int \dfrac{\ln x}{x}\, dx = \dfrac{1}{2}(\ln x)^2$. Therefore,

    $\displaystyle\int_{e^{1/p}}^{\infty} \dfrac{\ln x}{x^p}\, dx$ diverges when $p \le 1$ and converges when $p > 1$. It follows that the series converges only when $p > 1$.

16. The alternating series test (Theorem 11) implies that the series converges for every $p > 0$. (When $p > 0$, the terms of the series decrease in magnitude for $k > e^{1/p}$, approach zero, and alternate in sign.)

17. The series converges absolutely when $p > 1$ since $\displaystyle\sum_{k=2}^{\infty} \dfrac{\ln k}{k^p}$ converges only when $p > 1$.

18. When $0 < p \le 1$, $\displaystyle\sum_{k=2}^{\infty} \dfrac{\ln k}{k^p}$ diverges but $\displaystyle\sum_{k=2}^{\infty} (-1)^k \dfrac{\ln k}{k^p}$ converges. Therefore, the series converges conditionally when $0 < p \le 1$.

19. converges absolutely—$\displaystyle\sum_{j=1}^{\infty} \frac{1}{j^2}$ is a convergent p-series ($p = 2$). $\dfrac{3}{4} < \displaystyle\sum_{j=1}^{\infty} \frac{(-1)^{j+1}}{j^2} < 1.$

20. converges conditionally—$\displaystyle\sum_{k=1}^{\infty} \frac{1}{\sqrt{k}}$ is a divergent p-series ($p = 1/2$) but the terms of the

    series form a decreasing sequence and $\displaystyle\lim_{k\to\infty} \frac{1}{\sqrt{k}} = 0.$ $-1 < \displaystyle\sum_{k=1}^{\infty} \frac{(-1)^k}{\sqrt{k}} < -1 + \frac{1}{\sqrt{2}}.$

21. diverges—nth term test: $\displaystyle\lim_{n\to\infty} \frac{(-3)^n}{n^3}$ does not exist.

22. converges conditionally—$\displaystyle\sum_{k=4}^{\infty} \frac{\ln k}{k}$ diverges by the integral test but the terms of the series

    form a decreasing sequence and $\displaystyle\lim_{k\to\infty} \frac{\ln k}{k} = 0.$ $\dfrac{\ln 4}{4} - \dfrac{\ln 5}{5} < \displaystyle\sum_{k=4}^{\infty} (-1)^k \frac{\ln k}{k} < \frac{\ln 4}{4}.$

23. diverges—nth term test: $\displaystyle\lim_{k\to\infty} \frac{k}{2k+1} = \frac{1}{2} \neq 0.$

24. converges absolutely—Let $a_m = m^3/2^m$. Then $\displaystyle\lim_{m\to\infty} \frac{a_{m+1}}{a_m} = \lim_{m\to\infty} \frac{(m+1)^3}{2m^3} = \frac{1}{2} < 1$ so

    $\displaystyle\sum_{m=0}^{\infty} \frac{m^3}{2^m}$ converges by the ratio test. In addition, the terms of the series are decreasing in

    absolute value for all $m \geq 4$. Thus,

    $$-\frac{57}{32} = \sum_{m=0}^{5} (-1)^m a_m < \sum_{m=0}^{\infty} (-1)^m \frac{m^3}{2^m} < \sum_{m=0}^{4} (-1)^m a_m = \frac{17}{8}.$$

25. converges absolutely—$\displaystyle\lim_{j\to\infty} \frac{a_{j+1}}{a_j} = \lim_{j\to\infty} \frac{j+1}{(j^2+2j+1)\cdots(j^2+1)} = 0.$ The terms of the

    series are decreasing in absolute value for all $j \geq 1$. Thus,

    $$0 = \sum_{j=0}^{1} (-1)^j a_j < \sum_{j=0}^{\infty} (-1)^j a_j < \sum_{j=0}^{2} (-1)^j a_j = \frac{1}{12}.$$

26. converges conditionally—$\displaystyle\sum_{n=1}^{\infty} \frac{\arctan n}{n} > \frac{\pi}{4} \sum_{n=1}^{\infty} \frac{1}{n}.$ Thus,

    $$\frac{\pi}{4} - \frac{\arctan 2}{2} < \sum_{n=1}^{\infty} (-1)^{n+1} \frac{\arctan n}{n} < \frac{\pi}{4}.$$

27. Let $b_k = a_{k+10^9}$. The alternating series test can be used to show that $\displaystyle\sum_{k=1}^{\infty} (-1)^{k+1} b_k$

    converges. Since

$$\sum_{k=1}^{\infty}(-1)^{k+1}a_k = \sum_{k=1}^{10^9}(-1)^{k+1}a_k + \sum_{k=10^9+1}^{\infty}(-1)^{k+1}a_k = \sum_{k=1}^{10^9}(-1)^{k+1}a_k + \sum_{k=1}^{\infty}(-1)^{k+1}b_k, \text{ the}$$

series $\sum_{k=1}^{\infty}(-1)^{k+1}a_k$ converges.

28. The series converges absolutely because

$$1 + \frac{1}{2^3} + \frac{1}{3^2} + \frac{1}{4^3} + \frac{1}{5^2} + \frac{1}{6^3} + \frac{1}{7^2} + \frac{1}{8^3} + \cdots < \sum_{k=1}^{\infty}\frac{1}{k^2}.$$

NOTE: Theorem 11 cannot be used to show that the given series converges because the terms are *not* decreasing in magnitude.

29. Theorem 9 implies that $\sum_{j=1}^{\infty}(-1)^{j+1}b_j$ converges absolutely.

30. Since the terms of the series defining S are all positive, $S - \sum_{j=1}^{100}b_j \leq 0.005$ implies that $b_j \leq 0.005$ for all $j \geq 101$. Therefore, since $b_{j+1} \leq b_j$, Theorem 11 implies the desired result.

Alternatively,

$$\left|\sum_{j=1}^{\infty}(-1)^{j+1}b_j - \sum_{j=1}^{100}(-1)^{j+1}b_j\right| = \left|\sum_{j=101}^{\infty}(-1)^{j+1}b_j\right| \leq \sum_{j=101}^{\infty}b_j \leq 0.005.$$

31. $a_k = (-1)^k/\sqrt{k}$.

32. **No.** Since $|a_k| \geq a_k$, this would contradict Theorem 10.

33. (a) For $n = 1, 2, 3, \ldots,$ $S_{2n+2} = S_{2n} + c_{2n+1} - c_{2n+2} \geq S_{2n}$ since $c_{2n+1} - c_{2n+2} \geq 0.$

     (b) For $n = 1, 2, 3, \ldots,$ $S_{2n+1} = S_{2n-1} - c_{2n} + c_{2n+1} \leq S_{2n-1}$ since $c_{2n+1} - c_{2n} \leq 0.$

     (c) $S_{2m} = S_{2m-1} - c_{2m} \implies S_{2m} \leq S_{2m-1}$ since $c_{2m} \geq 0.$

     (d) It follows from parts (a)–(c) that $S_2 \leq S_{2m} \leq S_{2m-1} \leq S_1$ for $m = 1, 2, 3, \ldots.$ Thus, because the sequence of even partial sums is increasing and bounded above by S_1, it converges. Similarly, the sequence of odd partial sums converges because it is decreasing and bounded below by S_2.

     (e) $\lim_{m\to\infty}(S_{2m+1} - S_{2m}) = \lim_{m\to\infty}c_{2m+1} = 0.$ This implies that the limit of the sequence of even partial sums is the same as the limit of the sequence of odd partial sums.

(f) Since the sequence of even partial sums is increasing, $0 < S - S_{2m}$ is true. Also, since the sequence of odd partial sums is decreasing,

$S < S_{2m+1} = S_{2m} + c_{2m+1} \implies S - S_{2m} < c_{2m+1}$. Thus, $0 < S - S_{2m} < c_{2m+1}$.

Similarly, since the sequence of odd partial sums is decreasing, $0 < S_{2m+1} - S$. Also, since the sequence of even partial sums is increasing,

$S > S_{2m+2} = S_{2m+1} - c_{2m+2} \implies S - S_{2m+1} > -c_{2m+2}$. Thus,

$0 < S_{2m+1} - S < c_{2m+2}$.

§11.5   Power Series

1. $P_1(x) = x$, $P_2(x) = P_1(x) + x^2/2$, $P_4(x) = P_2(x) + x^3/3 + x^4/4$,
   $P_6(x) = P_4(x) + x^5/5 + x^6/6$, $P_8(x) = P_6(x) + x^7/7 + x^8/8$,
   $P_{10}(x) = P_8(x) + x^9/9 + x^{10}/10$.

3. $\displaystyle\lim_{j\to\infty}\left|\frac{(x/2)^{j+1}}{(x/2)^j}\right| = \lim_{j\to\infty}|x|/2 < 1$ when $|x| < 2$. Thus, the radius of convergence is $R = 2$.

4. $\left|\dfrac{a_{k+1}}{a_k}\right| = \dfrac{k}{3(k+1)} \cdot |x| < 1 \implies R = 3$.

5. $\displaystyle\lim_{k\to\infty}\left|\frac{\frac{x^{k+1}}{\sqrt{k+1}}}{\frac{x^k}{\sqrt{k}}}\right| = \lim_{k\to\infty}\frac{\sqrt{k}}{\sqrt{k+1}} \cdot |x| < 1$ when $|x| < 1$. Thus, the radius of convergence is
   $R = 1$.

6. $\displaystyle\lim_{n\to\infty}\left|\frac{x^n}{n! + n}\right| \leq \lim_{n\to\infty}\frac{|x|^n}{n!} = 0 \implies R = \infty$.

7. $\left|\dfrac{a_{n+1}}{a_n}\right| = |x - 2| < 1 \implies R = 1$; interval of convergence is $(1, 3)$.

8. $R = 1$; interval of convergence is $[2, 4]$.

9. $R = 1$; interval of convergence is $[-6, -4)$.

10. $R = 1$; interval of convergence is $[-2, 0)$.

11. The nth term test can be used to prove that the series diverges when $|x| \geq R$.

12. The alternating series test can be used to show that the series converges when $x = -R$.
    When $x = R$ the series becomes the harmonic series (i.e., it diverges).

13. When $x = -R$ the series becomes $\displaystyle\sum_{k=1}^{\infty}\frac{(-1)^k}{k^2}$, which converges absolutely.

14. $\displaystyle\sum_{k=1}^{\infty}\frac{(-x)^k}{kR^k} = \sum_{k=1}^{\infty}(-1)^k\frac{x^k}{kR^k}$.

15. $\displaystyle\sum_{k=1}^{\infty}\frac{x^k}{k4^k}$.

16. $\displaystyle\sum_{k=1}^{\infty}\frac{(-x)^k}{k3^k}$.

17. $\displaystyle\sum_{k=1}^{\infty} \frac{(x-2)^k}{k^2 3^k}.$

18. $\displaystyle\sum_{k=1}^{\infty} \frac{(x+2)^k}{2^k}.$

19. $\displaystyle\sum_{k=1}^{\infty} \frac{(12-x)^k}{k4^k}.$

20. $\displaystyle\sum_{k=1}^{\infty} \frac{(x+7)^k}{k4^k}.$

21. By definition, the radius of convergence of a power series (whose base point is zero) is the largest value of R such that the series converges for all x such that $|x| < R$.

22. This power series converges when $|x-1| < 2$ and diverges when $|x-1| > 2$. Thus, its radius of convergence is $R = 2$.

23. Let $z = x - 3$. Since $\displaystyle\sum_{k=0}^{\infty} a_k z^k$ converges only when $-2 < z \le 2$, $\displaystyle\sum_{k=0}^{\infty} a_k (x-3)^k$ converges only when $1 < x \le 5$.

24. The power series $\displaystyle\sum_{k=0}^{\infty} a_k (x+1)^k$ converges only when $-2 < x + 1 \le 2$. Thus, its interval of convergence is $(-3, 1]$.

25. (a) $R = 14$.

    (b) $b = 3$.

26. $\displaystyle\sum_{n=1}^{\infty} a_n |x|^n < \sum_{n=1}^{\infty} a_n \cdot 1^n = \sum_{n=1}^{\infty} a_n$, which converges.

27. Cannot.

    This power series has an interval of convergence centered at $x = 0$. Since it converges for $x = -3$, it must converge if $-3 \le x < 3$. Since it diverges for $x = 7$, it must diverge if $x < -7$ or $x \ge 7$. The information given is insufficient to conclude anything about the convergence of the power series on the intervals $[-7, -3)$ and $[3, 7)$.

28. may (see discussion in solution to #27).

29. may (see discussion in solution to #27).

30. cannot (see discussion in solution to #27).

31. may (see discussion in solution to #27).

32. may (see discussion in solution to #27).

33. May. The information given implies that the power series converges on the interval $[-7, 3)$, diverges when $x \geq 7$, and diverges when $x < -11$. It does not imply anything about convergence or divergence on the intervals $[-11, -7)$ and $[3, 7)$.

34. must (see discussion in solution to #33).

35. may (see discussion in solution to #33).

36. may (see discussion in solution to #33).

37. may (see discussion in solution to #33).

38. cannot (see discussion in solution to #33).

39. The series $\displaystyle\sum_{n=0}^{\infty} \frac{2 \cdot 10^n}{3^n + 5}$ diverges by the ratio test—the limit of the ratio of successive terms of the series is $10/3 > 1$.

40. The power series defining f converges when $-3 < x < 3$. Thus, only 0.5 and 1.5 are in the domain of f.

41. $\displaystyle f(1) - \sum_{n=0}^{N} \frac{2}{3^n + 5} = \sum_{n=N+1}^{\infty} \frac{2}{3^n + 5} < \sum_{n=N+1}^{\infty} \frac{2}{3^n} = \frac{1}{3^{N+1}} \sum_{n=0}^{\infty} \frac{2}{3^n} = \frac{2}{3^{N+1}} \cdot \frac{3}{2} = \frac{1}{3^N}$. Thus,

    since $3^{-5} < 0.01$, $\displaystyle\sum_{n=0}^{5} \frac{2}{3^n + 5} = \frac{367273}{447888} \approx 0.82001$ approximates $f(1)$ within 0.01.

42. The value of $f(-1.5)$ is defined by an alternating series. Since $\left| \dfrac{2(-1.5)^8}{3^8 + 5} \right| < 0.01$,

    $\displaystyle f(-1.5) \approx \sum_{n=0}^{7} \frac{2(-1.5)^n}{3^n + 5} \approx 0.14076$.

43. The series defining h converges for all values of x.

44. $\displaystyle h(0) = \sum_{k=0}^{\infty} \frac{(-2)^k}{k! + k^3}$. Since this is an alternating series whose terms decrease in magnitude

    for all $k \geq 1$ and since $\dfrac{2^9}{9! + 9^3} < 0.005$, $\displaystyle h(0) \approx \sum_{k=0}^{8} \frac{(-2)^k}{k! + k^3} = \frac{1825808722}{7031839815} \approx 0.25965$.

45. The value of $h(1)$ is defined by an alternating series. Since $\left| \dfrac{(-1)^5}{5! + 5^3} \right| < 0.01$,

$$h(1) \approx \sum_{k=0}^{4} \frac{(-1)^k}{k! + k^3} \approx 0.58106.$$

46. $h(3) = \displaystyle\sum_{k=0}^{\infty} \frac{1}{k! + k^3}$. Since $\dfrac{1}{k! + k^3} < \dfrac{1}{k!}$ for all $k \geq 1$,

$$\sum_{k=6}^{\infty} \frac{1}{k! + k^3} < \sum_{k=6}^{\infty} \frac{1}{k!} = e - \frac{163}{60} \approx 0.0016152 < 0.005.$$

Therefore, $h(3) \approx \displaystyle\sum_{k=0}^{5} \frac{1}{k! + k^3} = \frac{9677}{5880} \approx 1.6457.$

47. The domain of g is $[-9, 1]$.

48. $g(0) - \displaystyle\sum_{n=1}^{N} \frac{4^n}{n^3 5^n} = \sum_{n=N+1}^{\infty} \frac{4^n}{n^3 5^n} < \int_{N}^{\infty} \frac{dx}{x^3} = \frac{1}{2N^2} \leq 0.005$ when $N \geq 10$. Thus,

$$g(0) \approx \sum_{n=1}^{10} \frac{4^n}{n^3 5^n} = \frac{277892997449134}{305233154296875} \approx 0.91043.$$

49. $g(1) = \displaystyle\sum_{n=1}^{\infty} \frac{1}{n^3}$. Now, the integral test implies that $R_{10} \leq \displaystyle\int_{10}^{\infty} \frac{dx}{x^3} = 0.005$ so

$$g(1) \approx \sum_{n=1}^{10} \frac{1}{n^3} \approx 1.19753.$$

50. The approximation $g(-5) \approx -\dfrac{1}{5}$ is correct within 0.005 because the series defining $g(-5)$ is a monotonically decreasing, alternating series. Since the magnitude of the second term in the series is 0.005, the error made by approximating the series by its first term is smaller than 0.005.

§11.6 Power Series as Functions

1. Since $\left|\dfrac{a_{k+1}}{a_k}\right| = \dfrac{|x|}{2}$, the radius of convergence is 2.

2. Since $\left|\dfrac{a_{k+1}}{a_k}\right| = \dfrac{(k+1)|x|}{2k}$, the radius of convergence is 2.

3. Since $\left|\dfrac{a_{k+1}}{a_k}\right| = \dfrac{(k+1)|x|}{2(k+2)}$, the radius of convergence is 2.

4. The power series representation of $f(x) = \ln(1+x)$ converges on the interval $(-1, 1]$. The power series representation for $f'(x) = 1/(1+x)$ converges on the interval $(-1, 1)$.

5. $f(x) = \dfrac{x^2}{1+x} = x^2 \displaystyle\sum_{k=0}^{\infty} (-1)^k x^k = \sum_{k=0}^{\infty} (-1)^k x^{k+2}$ [Substitute $u = -x$ into the power series representation of $(1-u)^{-1}$.]

6. $f(x) = \left(1 - x^2\right)^{-1} = \displaystyle\sum_{k=0}^{\infty} x^{2k}$ [Substitute $u = x^2$ into the power series representation of $(1-u)^{-1}$.]

7. $f(x) = (1+x)^{-2} = -\dfrac{d}{dx}\left((1+x)^{-1}\right) = -\dfrac{d}{dx}\left(\displaystyle\sum_{k=0}^{\infty} (-x)^k\right) = \sum_{k=1}^{\infty} k(-x)^{k-1}$.

8. $f(x) = \dfrac{x}{1 - x^4} = x \displaystyle\sum_{k=0}^{\infty} x^{4k} = \sum_{k=0}^{\infty} x^{4k+1}$.

9. $\arctan(2x) = \displaystyle\sum_{k=0}^{\infty} (-1)^k \dfrac{(2x)^{2k+1}}{2k+1}$; $R = \dfrac{1}{2}$; $P = 2x - \dfrac{8x^3}{3} + \dfrac{32x^5}{5} - \dfrac{128x^7}{7} + \dfrac{512x^9}{9}$.

10. $\cos(x^2) = \displaystyle\sum_{k=0}^{\infty} (-1)^k \dfrac{x^{4k}}{(2k)!}$; $R = \infty$; $P = 1 - \dfrac{x^4}{2} + \dfrac{x^8}{24} - \dfrac{x^{12}}{720} + \dfrac{x^{16}}{40320}$.

11. $x^2 \sin x = \displaystyle\sum_{k=0}^{\infty} (-1)^k \dfrac{x^{2k+3}}{(2k+1)!}$; $R = \infty$; $P = x^3 - \dfrac{x^5}{6} + \dfrac{x^7}{120} - \dfrac{x^9}{5040} + \dfrac{x^{11}}{362880}$.

12. $\ln\left(1 + \sqrt[3]{x}\right) = \displaystyle\sum_{k=1}^{\infty} (-1)^{k+1} \dfrac{x^{k/3}}{k}$; $R = 1$; $P = x^{1/3} - \dfrac{x^{2/3}}{2} + \dfrac{x}{3} - \dfrac{x^{4/3}}{4} + \dfrac{x^{5/3}}{5}$.

13. $1/\sqrt{e} = e^{-1/2} = \displaystyle\sum_{k=0}^{\infty} \dfrac{(-1/2)^k}{k!}$. Since $1/(2^4 \cdot 4!) < 0.005$, $\displaystyle\sum_{k=0}^{3} \dfrac{(-1/2)^k}{k!} = \dfrac{29}{48} \approx 0.60417$ approximates $1/\sqrt{e}$ within 0.005.

14. $\int_0^{0.2} xe^{-x^3}\, dx = \int_0^{0.2} \left(\sum_{k=0}^{\infty} \frac{(-x)^{3k+1}}{k!} \right) dx =$

$\sum_{k=0}^{\infty} \frac{(-0.2)^{3k+2}}{(3k+2)\, k!}$. Since $(-0.2)^8/(8 \cdot 2!) < 10^{-5}$,

$\int_0^{0.2} xe^{-x^3}\, dx \approx \sum_{k=0}^{1} \frac{(-0.2)^{3k+2}}{(3k+2)\, k!} = \frac{623}{31250} = 0.019936$ approximates the given definite

integral within 10^{-5}.

15. $(\sin x - x)^3 = \left(\sum_{k=1}^{\infty} (-1)^k \frac{x^{2k+1}}{(2k+1)!} \right)^3 = \left(-\frac{x^3}{3!} + \frac{x^5}{5!} \mp \cdots \right)^3 = -\frac{x^9}{(3!)^3} + \frac{x^{11}}{1440} \mp \cdots.$

$(1 - \cos x)^4 = \left(\sum_{k=1}^{\infty} (-1)^{k+1} \frac{x^{2k}}{(2k)!} \right)^4 = \left(\frac{x^2}{2!} - \frac{x^4}{4!} \pm \cdots \right)^4 = \frac{x^8}{2^4} - \frac{x^{10}}{48} \pm \cdots.$

Therefore,

$$\lim_{x \to 0} \frac{(\sin x - x)^3}{x(1 - \cos x)^4} = \lim_{x \to 0} \frac{-x^9/216 + x^{11}/1440 \mp \cdots}{x \cdot (x^8/16 - x^{10}/48 \pm \cdots)}$$

$$= \lim_{x \to 0} \frac{-1/216 + x^2/1440 \mp \cdots}{1/16 - x^2/48 \pm \cdots} = -\frac{2}{27}.$$

16. $x - \sin x = \sum_{k=1}^{\infty} (-1)^{k+1} \frac{x^{2k+1}}{(2k+1)!} = \frac{x^3}{3!} - \frac{x^5}{5!} \pm \cdots = x^3 \left(\frac{1}{3!} - \frac{x^2}{5!} \pm \cdots \right)$

$(x \sin x)^{3/2} = \left(\sum_{k=0}^{\infty} (-1)^k \frac{x^{2k+2}}{(2k+1)!} \right)^{3/2} = \left(x^2 - \frac{x^4}{3!} \pm \cdots \right)^{3/2}$

$= \left(x^6 - \frac{1}{2} x^8 \pm \cdots \right)^{1/2} = x^3 \left(1 - \frac{1}{2} x^2 \pm \cdots \right)^{1/2}.$

Thus,

$$\lim_{x \to 0^+} \frac{x - \sin x}{(x \sin x)^{3/2}} = \lim_{x \to 0^+} \frac{x^3 \left(\frac{1}{3!} - \frac{x^2}{5!} \pm \cdots \right)}{x^3 \left(1 - \frac{1}{2} x^2 \pm \cdots \right)^{1/2}} = \frac{1}{6}.$$

17. $\frac{1}{2+x} = \frac{1}{2} \left(\frac{1}{1 + (x/2)} \right) = \frac{1}{2} \sum_{k=0}^{\infty} (-1)^k \left(\frac{x}{2} \right)^k; R = 2.$

18. $\sin\left(\sqrt{x} \right) = \sum_{k=0}^{\infty} (-1)^k \frac{x^{(2k+1)/2}}{(2k+1)!}; R = \infty.$

19. $\sin x + \cos x = \sum_{k=0}^{\infty} (-1)^k \left(\frac{x^{2k}}{(2k)!} + \frac{x^{2k+1}}{(2k+1)!} \right); R = \infty.$

20. $\ln\left(1+x^2\right) = \sum_{k=0}^{\infty}(-1)^k\dfrac{x^{2k+2}}{k+1}; \ R = 1.$

21. $(x^2 - 1)\sin x = \left(x^2 - 1\right)\sum_{k=0}^{\infty}(-1)^k\dfrac{x^{2k+1}}{(2k+1)!}$

$$= -x + \sum_{k=1}^{\infty}(-1)^{k+1}\dfrac{\big((2k+1)(2k)+1\big)x^{2k+1}}{(2k+1)!}$$

$$= \sum_{k=0}^{\infty}(-1)^{k+1}\dfrac{(4k^2+2k+1)x^{2k+1}}{(2k+1)!}; \ R = \infty.$$

22. $\ln\left(\dfrac{1+x}{1-x}\right) = \ln(1+x) - \ln(1-x) = 2\sum_{k=0}^{\infty}\dfrac{x^{2k+1}}{2k+1}; \ R = 1.$

23. $y = e^x = \sum_{k=0}^{\infty}\dfrac{x^k}{k!} \implies y' = \sum_{k=0}^{\infty}\dfrac{kx^{k-1}}{k!} = \sum_{k=1}^{\infty}\dfrac{x^{k-1}}{(k-1)!} = \sum_{k=0}^{\infty}\dfrac{x^k}{k!} = y.$

24. $y = 2e^x = \sum_{k=0}^{\infty}\dfrac{2x^k}{k!} \implies y' = \sum_{k=0}^{\infty}\dfrac{2kx^{k-1}}{k!} = \sum_{k=1}^{\infty}\dfrac{2x^{k-1}}{(k-1)!} = \sum_{k=0}^{\infty}\dfrac{2x^k}{k!} = y.$

25. $y = e^{3x} = \sum_{k=0}^{\infty}\dfrac{(3x)^k}{k!} \implies y' = 3\sum_{k=0}^{\infty}\dfrac{k(3x)^{k-1}}{k!} = 3\sum_{k=1}^{\infty}\dfrac{(3x)^{k-1}}{(k-1)!} = 3\sum_{k=0}^{\infty}\dfrac{(3x)^k}{k!} = 3y.$

26. $y = \sin x = \sum_{k=0}^{\infty}(-1)^k\dfrac{x^{2k+1}}{(2k+1)!}$ and $y' = \sum_{k=0}^{\infty}(-1)^k\dfrac{(2k+1)x^{2k}}{(2k+1)!} = \sum_{k=0}^{\infty}(-1)^k\dfrac{x^{2k}}{(2k)!}$, so

$$y'' = \sum_{k=0}^{\infty}(-1)^k\dfrac{(2k)x^{2k-1}}{(2k)!} = \sum_{k=1}^{\infty}(-1)^k\dfrac{x^{2k-1}}{(2k-1)!} = \sum_{k=0}^{\infty}(-1)^{k+1}\dfrac{x^{2k+1}}{(2k+1)!} = -y.$$

27. $y = (1-x)^{-1} = \sum_{k=0}^{\infty}x^k \implies y' = \sum_{k=0}^{\infty}kx^{k-1} = \sum_{k=0}^{\infty}(k+1)x^k = \left(\sum_{k=0}^{\infty}x^k\right)^2 = y^2.$

28.   (a) $1 + \big(f(x)\big)^2 = 1 + \tan^2 x = \sec^2 x = f'(x).$

    (b) Since $f(0) = 0$, we assume that $f(x) = \sum_{k=1}^{\infty}a_k x^k$. Inserting this power series into the

    identity from part (a):

    $a_1 + 2a_2 x + 3a_3 x^2 + \cdots = 1 + a_1^2 x^2 + 2a_1 a_2 x^3 + (2a_1 a_3 + a_2)x^4 + (2a_1 a_4 + 2a_2 a_3)$
    $x^5 + (2a_1 a_5 + 2a_2 a_4 + a_3^2)x^6 + \cdots$. Equating powers of x on both sides and solving,
    we find that $a_1 = 1, a_2 = 0, a_3 = 1/3, a_4 = 0, a_5 = 2/15, a_6 = 0$, and $a_7 = 17/315.$

    Thus, $\tan x = x + \dfrac{x^3}{3} + \dfrac{2x^5}{15} + \dfrac{17x^7}{315} + \cdots.$

29. $\dfrac{\sin x}{x} = \displaystyle\sum_{k=0}^{\infty}(-1)^k \dfrac{x^{2k}}{(2k+1)!} = 1 - x^2/3! + x^4/5! - x^6/7! + \cdots \implies \displaystyle\lim_{x \to 0} \dfrac{\sin x}{x} = 1.$

30. $\dfrac{e^x - 1}{x} = x^{-1}\left(\displaystyle\sum_{k=0}^{\infty}\dfrac{x^k}{k!} - 1\right) = \displaystyle\sum_{k=1}^{\infty}\dfrac{x^{k-1}}{k!} \implies \displaystyle\lim_{x \to 0}\dfrac{e^x - 1}{x} = 1.$

31. $\dfrac{1 - \cos x}{x^2} = x^{-2}\left(1 - \displaystyle\sum_{k=0}^{\infty}(-1)^k\dfrac{x^{2k}}{(2k)!}\right) = \displaystyle\sum_{k=0}^{\infty}(-1)^k\dfrac{x^{2k}}{(2k+2)!} \implies \displaystyle\lim_{x \to 0}\dfrac{1 - \cos x}{x^2} = \dfrac{1}{2}.$

32. $\dfrac{\arctan x}{x} = x^{-1}\displaystyle\sum_{k=0}^{\infty}(-1)^k\dfrac{x^{2k+1}}{2k+1} = \displaystyle\sum_{k=0}^{\infty}(-1)^k\dfrac{x^{2k}}{2k+1} \implies \displaystyle\lim_{x \to 0}\dfrac{\arctan x}{x} = 1.$

33. $\dfrac{\ln(1 + x) - x}{x^2} = \displaystyle\sum_{k=0}^{\infty}(-1)^{k+1}\dfrac{x^k}{k+2} \implies \displaystyle\lim_{x \to 0}\dfrac{\ln(1 + x) - x}{x^2} = -\dfrac{1}{2}.$

34. $\dfrac{1 - \cos^2 x}{x} = \dfrac{\frac{1}{2} - \frac{1}{2}\cos(2x)}{x} = \displaystyle\sum_{k=1}^{\infty}(-1)^{k+1}\dfrac{(2x)^{2k-1}}{(2k)!} \implies \displaystyle\lim_{x \to 0}\dfrac{1 - \cos^2 x}{x} = 0.$

35. Let $f(x) = 1/(1 - x) = \displaystyle\sum_{n=0}^{\infty} x^n$ if $|x| < 1$. Then

$$f'(x) = 1/(1 - x)^2 = \sum_{n=1}^{\infty}nx^{n-1} = \dfrac{1}{x}\sum_{n=1}^{\infty}nx^n \text{ if } |x| < 1 \text{ and } x \neq 0. \text{ Therefore,}$$

$$\sum_{n=1}^{\infty}\dfrac{n}{2^n} = (1/2)f'(1/2) = 2.$$

36. $\dfrac{1}{x - 1} = \dfrac{x}{x - 1} - 1 = \dfrac{1}{1 - (1/x)} - 1 = \displaystyle\sum_{k=0}^{\infty}\left(\dfrac{1}{x}\right)^k - 1 = \displaystyle\sum_{k=1}^{\infty}\dfrac{1}{x^k}$ since $|1/x| < 1$ if $|x| > 1$.

37. (a) Integrating term by term, $\dfrac{1}{1 - x} = \displaystyle\sum_{k=0}^{\infty}x^k \implies -\ln|1 - x| = \displaystyle\sum_{k=1}^{\infty}\dfrac{x^k}{k}.$

   (b) The series converges on the interval $[-1, 1)$.

   (c) When $x = 1/2$, part (a) implies that $-\ln(1/2) = \ln 2 = \displaystyle\sum_{k=1}^{\infty}\dfrac{1}{k\,2^k}$. Since the terms of

   this series are all positive, the partial sums $S_N = \displaystyle\sum_{k=1}^{N}\dfrac{1}{k\,2^k}$ form an increasing sequence

   that is bounded above by the sum of the series. Thus, $\ln 2 - S_N > 0$.

$$\ln 2 - S_N = \sum_{k=1}^{\infty} \frac{1}{k \, 2^k} - \sum_{k=1}^{N} \frac{1}{k \, 2^k} = \sum_{k=N+1}^{\infty} \frac{1}{k \, 2^k} \le \frac{1}{N+1} \sum_{k=N+1}^{\infty} \frac{1}{2^k}$$

$$= \frac{1}{(N+1)2^{N+1}} \sum_{k=0}^{\infty} \frac{1}{2^k} = \frac{1}{(N+1)2^N}.$$

38. The power series representation of $\ln(1+x)$ is a convergent alternating series if $0 < x < 1$:

$$\ln(1+x) = \sum_{k=1}^{\infty} \frac{(-1)^{k+1} x^k}{k}.$$

Because the partial sums of a convergent alternating series alternately overestimate and underestimate the limit, the first two partial sums bound the value of $\ln(1+x)$.

39. The power series representations of each of the three functions are alternating series if $x > 0$, so the following inequalities are valid if $0 < x < 1$:
$x - x^2/2 < \ln(1+x) < x - x^2/2 + x^3/3$; $x - x^3/3! < \sin x < x$; and
$x^2/2! - x^4/4! < 1 - \cos x < x^2/2!$.

Since $(x - x^3/3!) - (x - x^2/2 + x^3/3) = x^2(1-x)/2 > 0$ if $0 < x < 1$, the lower bound on $\sin x$ is greater than the upper bound on $\ln(1+x)$, so $\ln(1+x) < \sin x$ if $0 < x < 1$.
Also, since $(x - x^2/2) - x^2/2! = x(1-x) > 0$ if $0 < x < 1$, the lower bound on $\ln(1+x)$ is greater than the upper bound on $1 - \cos x$, so $1 - \cos x < \ln(1+x)$ if $0 < x < 1$.

40. (a) $\dfrac{1}{1+x^4} = \sum_{k=0}^{\infty} (-1)^k x^{4k}.$

    (b) $(-1, 1)$.

    (c) $\displaystyle\int f(x)\,dx = \int \frac{dx}{1+x^4} = \int \left(\sum_{k=0}^{\infty} (-1)^k x^{4k} \right) dx = \sum_{k=0}^{\infty} (-1)^k \frac{x^{4k+1}}{4k+1}.$ Therefore,

    $$\int_0^{0.5} f(x)\,dx = \sum_{k=0}^{\infty} (-1)^k \frac{(0.5)^{4k+1}}{4k+1}.$$

    Since $\dfrac{(0.5)^9}{9} < 0.001$, $\displaystyle\int_0^{0.5} f(x)\,dx \approx \sum_{k=0}^{1} (-1)^k \frac{(0.5)^{4k+1}}{4k+1} = \frac{79}{160} = 0.49375.$

41. (a) $\displaystyle\int e^{-x^2}\,dx = \int \left(\sum_{k=0}^{\infty} \frac{(-x^2)^k}{k!} \right) dx = \sum_{k=0}^{\infty} \frac{(-1)^k}{(2k+1) \cdot k!} x^{2k+1}.$

    (b) The approximation $\displaystyle\int_0^1 e^{-x^2}\,dx \approx \sum_{k=0}^{3} \frac{(-1)^k}{(2k+1) \cdot k!} = \frac{26}{35}$ has the desired accuracy

    because $\dfrac{1}{9 \cdot 4!} < 0.005$.

42. $\displaystyle\int_0^1 \cos\left(x^2\right)\,dx = \int_0^1 \left(\sum_{k=0}^{\infty} \frac{(-1)^k x^{4k}}{(2k)!}\right)\,dx = \sum_{k=0}^{\infty} \frac{(-1)^k x^{4k+1}}{(4k+1)\cdot(2k)!}\Bigg|_0^1$

$\displaystyle = \sum_{k=0}^{\infty} \frac{(-1)^k}{(4k+1)\cdot(2k)!} \approx \sum_{k=0}^{1} \frac{(-1)^k}{(4k+1)\cdot(2k)!} = \frac{9}{10}$ within 0.005.

43. $\displaystyle\int_0^1 \sqrt{x}\,\sin x\,dx = \int_0^1 \left(\sum_{k=0}^{\infty} (-1)^k \frac{x^{(4k+3)/2}}{(2k+1)!}\right)\,dx = \sum_{k=0}^{\infty} \frac{(-1)^k 2}{(4k+5)(2k+1)!}$

$\displaystyle \approx \sum_{k=0}^{2} \frac{(-1)^k 2}{(4k+5)(2k+1)!} = \frac{2557}{7020} \approx 0.36425.$

44. $\displaystyle\int_0^1 \frac{\sin x}{x}\,dx = \int_0^1 \left(\sum_{k=0}^{\infty} \frac{(-1)^k x^{2k}}{(2k+1)!}\right)\,dx$

$\displaystyle = \sum_{k=0}^{\infty} \frac{(-1)^k}{(2k+1)!} \int_0^1 x^{2k}\,dx$

$\displaystyle = \sum_{k=0}^{\infty} \frac{(-1)^k}{(2k+1)(2k+1)!}.$

Since $\dfrac{1}{7\cdot 7!} < 5 \times 10^{-5}$, $\displaystyle\int_0^1 \frac{\sin x}{x}\,dx \approx \sum_{k=0}^{2} \frac{(-1)^k}{(2k+1)(2k+1)!} = \frac{1703}{1800} \approx 0.94611.$

45. $\displaystyle e^{2x}\ln\left(1+x^3\right) = \left(\sum_{k=0}^{\infty} \frac{(2x)^k}{k!}\right)\left(\sum_{k=1}^{\infty}(-1)^{k+1}\frac{x^{3k}}{k}\right) = x^3 + 2x^4 + 2x^5 + \frac{5}{6}x^6 + \cdots.$

46. $\displaystyle \arctan x\,\sin(4x) = \left(\sum_{k=0}^{\infty}(-1)^k \frac{x^{2k+1}}{2k+1}\right)\left(\sum_{k=0}^{\infty}(-1)^k \frac{(4x)^{2k+1}}{(2k+1)!}\right)$

$\displaystyle = 4x^2 - 12x^4 + \frac{116}{9}x^6 - \frac{44}{5}x^8 + \cdots.$

47. $\displaystyle \frac{e^x}{1-x} = \left(\sum_{k=0}^{\infty} \frac{x^k}{k!}\right)\left(\sum_{k=0}^{\infty} x^k\right) = 1 + 2x + \frac{5}{2}x^2 + \frac{8}{3}x^3 + \cdots.$

48. $\displaystyle \frac{\cos x}{1+x^2} = \left(\sum_{k=0}^{\infty}(-1)^k \frac{x^{2k}}{(2k)!}\right)\left(\sum_{k=0}^{\infty}(-1)^k x^{2k}\right) = 1 - \frac{3}{2}x^2 + \frac{37}{24}x^4 - \frac{1111}{720}x^6 + \cdots.$

49. $e^{\sin x} = \displaystyle\sum_{k=0}^{\infty} \frac{(\sin x)^k}{k!} = 1 + \sin x + \frac{\sin^2 x}{2!} + \frac{\sin^3 x}{3!} + \frac{\sin^4 x}{4!} + \cdots$

$= 1 + \left(x - x^3/3! + \cdots\right) + \frac{1}{2}\left(x - x^3/3! + \cdots\right)^2 + \frac{1}{3!}\left(x - \cdots\right)^3 + \frac{1}{4!}\left(x - \cdots\right)^4$

$= 1 + x + x^2/2 - x^4/8 + \cdots .$

50. $\ln(1 + \sin x) = \displaystyle\sum_{k=1}^{\infty} (-1)^{k+1} \frac{(\sin x)^k}{k}$

$= \left(x - x^3/6 + \cdots\right) - \frac{1}{2}\left(x^2 - x^4/3 + \cdots\right) + \frac{1}{3}\left(x^3 + \cdots\right) - \frac{1}{4}\left(x^4 + \cdots\right)$

$= x - x^2/2 + x^3/6 - x^4/12 + \cdots .$

51. $\displaystyle\sum_{k=1}^{\infty} k x^{k-1} = \left(\frac{1}{1-x}\right)' = \frac{1}{(1-x)^2}.$

52. $\displaystyle\sum_{k=0}^{\infty} \frac{x^k}{(k+1)!} = \frac{e^x - 1}{x}.$

53. $\displaystyle\sum_{k=1}^{\infty} (-1)^{k+1} x^k = 1 + \sum_{k=0}^{\infty} (-1)^{k+1} x^k = 1 - \sum_{k=0}^{\infty} (-1)^k x^k = 1 - \frac{1}{1+x} = \frac{x}{1+x}.$

54. $\displaystyle\sum_{k=1}^{\infty} \frac{(2x)^k}{k} = -\ln(1 - 2x).$

NOTE: $-\ln(1-x) = \displaystyle\int (1-x)^{-1}\, dx = \int \left(\sum_{k=0}^{\infty} x^k\right) dx = \sum_{k=1}^{\infty} \frac{x^k}{k}.$

§11.7 Taylor Series

1. (a) $f'(x) = \sqrt{x}e^{-x}$, $f''(x) = e^{-x}\left(\frac{1}{2\sqrt{x}} - \sqrt{x}\right)$, and $f'''(x) = e^{-x}\left(\sqrt{x} - \frac{1}{\sqrt{x}} - \frac{1}{4x^{3/2}}\right)$.

Therefore, $f(3) = 0$, $f'(3) = \sqrt{3}e^{-3}$, $f''(3) = -\frac{5\sqrt{3}}{6}e^{-3}$, and $f'''(3) = \frac{23\sqrt{3}}{36}e^{-3}$. It follows from Taylor's Theorem that

$$f(x) \approx \sqrt{3}e^{-3}(x-3) - \frac{5}{12}\sqrt{3}e^{-3}(x-3)^2 + \frac{23}{216}\sqrt{3}e^{-3}(x-3)^3.$$

(b) If $3 \le x \le 3.5$, $\left|f^{(4)}(x)\right| \le 0.035$, so the approximation error is bounded by

$$\frac{0.035 \cdot (0.5)^4}{4!} \approx 9.1146 \times 10^{-5}.$$

2. (a) $f(0) = 1$, $f'(0) = 1/2$, and $f''(0) = -1/4$ so $\sqrt{1+x} = 1 + \frac{x}{2} - \frac{x^2}{4 \cdot 2!} + \cdots$.

(b) The magnitude of the approximation error is bounded by $\frac{K_3}{3!} = \frac{3/8}{3!} = \frac{1}{16} \approx 0.0625$.

3. The Maclaurin series representation of f is

$$f(x) = \sum_{k=0}^{\infty} \frac{(2x)^k}{k!} = \sum_{k=0}^{\infty} \frac{f(k)(0)}{k!}x^k.$$

Thus, the coefficient of x^{100} in the Maclaurin series representation of f is $2^{100}/100!$.

4. (a) Using the fact that the Maclaurin series for $(1 - u)^{-1}$ is $\displaystyle\sum_{k=0}^{\infty} u^k$, the Maclaurin series for

f is $\displaystyle\sum_{k=0}^{\infty} x^{3k+1}$.

(b) The interval of convergence of the power series for f is $(-1, 1)$.

(c) Differentiating the power series for f term by term twice,

$$f''(x) = \sum_{k=1}^{\infty} (3k+1)(3k)x^{3k-1}.$$

(d) Integrating the power series for f term by term, $\displaystyle\int_0^x f(t)\,dt = \sum_{k=0}^{\infty} \frac{x^{3k+2}}{3k+2}$.

5. (a) $\dfrac{1}{2+x} = \dfrac{1}{2} \cdot \dfrac{1}{1+(x/2)} = \dfrac{1}{2} \cdot \dfrac{1}{1-(-x/2)} = \dfrac{1}{2}\sum_{k=0}^{\infty}\left(\dfrac{-x}{2}\right)^k = \sum_{k=0}^{\infty}\dfrac{(-1)^k x^k}{2^{k+1}}$.

(b) The coefficient of x^{259} in the Maclaurin series representation of $f(x)$ is $f^{(259)}(0)/259!$. Thus,
$$f^{(259)}(0) = -259!/2^{260}.$$

6. (a) $f(1.5) \approx f(1) + f'(1) \cdot (1.5 - 1) + \dfrac{f''(1)}{2}(1.5 - 1)^2$

$$= 1 + 2 \cdot (0.5) + \frac{1}{4}(0.5)^2 = \frac{33}{16} = 2.0625.$$

(b) Since $f'''(x) = -\dfrac{3x^2}{(1 + x^3)^2}$, $K_3 = 3/4$. Therefore, the approximation error is less than

or equal to $\dfrac{K_3}{3!}(1.5 - 1)^3 = \dfrac{1}{64} = 0.015625$.

7. $K_{n+1} = e^x$ so $\left| e^x - P_n(x) \right| \leq \dfrac{e^x \cdot |x|^{n+1}}{(n+1)!} \to 0$ as $n \to \infty$.

8. When $0 \leq x < 1$, $K_{n+1} = (n+1)!$ so $\left| \dfrac{1}{1+x} - P_n(x) \right| \leq \dfrac{(n+1)! \cdot |x|^{n+1}}{(n+1)!} = |x|^{n+1} \to 0$ as $n \to \infty$.

When $-1/2 < x < 0$, $K_{n+1} = 2^{n+2}(n+1)!$ so
$$\left| \frac{1}{1+x} - P_n(x) \right| \leq \frac{2^{n+2}(n+1)! \cdot |x|^{n+1}}{(n+1)!} = 2^{n+2}|x|^{n+1} \to 0 \text{ as } n \to \infty.$$

9. (a) If $\left| f^{(n)}(x) \right| \leq n$ for all $n \geq 1$, then $K_{n+1} \leq n+1$ and

$$|f(x) - P_n(x)| \leq \frac{(n+1)|x|^{n+1}}{(n+1)!} \to 0 \text{ as } n \to \infty.$$

(b) Yes, because $\displaystyle\lim_{n \to \infty} \frac{2^{n+1}|x|^{n+1}}{(n+1)!} = 0$ for all x.

10. The Maclaurin series representation of f is $f(x) = \displaystyle\sum_{k=0}^{\infty} \frac{f^{(k)}(0)}{k!}x^k = \sum_{k=0}^{\infty} (-1)^k \frac{1}{(2k+2)!}x^{2k}$.

Thus, $f^{(100)}(0)/100! = 1/102! \implies f^{(100)}(0) = 100!/102! = 1/(102 \cdot 101) = 1/10302$.

11. (a) $f(x) = x^{-1} \sin x = x^{-1} \displaystyle\sum_{k=0}^{\infty} \frac{(-1)^k x^{2k+1}}{(2k+1)!} = \sum_{k=0}^{\infty} \frac{(-1)^k x^{2k}}{(2k+1)!}$.

(b) The power series in part (a) converges for values of x in the interval $(-\infty, \infty)$.

(c) $f'''(x) = \displaystyle\sum_{k=2}^{\infty} \frac{(-1)^k (2k)(2k-1)(2k-2)x^{2k-3}}{(2k+1)!}$ for any $x \in (-\infty, \infty)$. Since $f'''(1)$ is

represented by an alternating series,
$$\left| f'''(1) - \sum_{k=2}^{3} \frac{(-1)^k (2k)(2k-1)(2k-2)}{(2k+1)!} \right| < \frac{8 \cdot 7 \cdot 6}{9!} = \frac{1}{1080} < 0.005;$$

$$f'''(1) \approx \frac{37}{210} \approx 0.17619.$$

12. (a) Using the definition of the derivative: $f'(0) = \lim\limits_{h \to 0} \dfrac{f(h) - f(0)}{h} = \lim\limits_{h \to 0} \dfrac{e^{-1/h^2}}{h}$. Now,

letting $x = 1/h$, $\lim\limits_{h \to 0^+} \dfrac{e^{-1/h^2}}{h} = \lim\limits_{x \to \infty} xe^{-x^2} = 0$ (by l'Hôpital's rule). Similarly,

$\lim\limits_{h \to 0^-} \dfrac{e^{-1/h^2}}{h} = \lim\limits_{x \to -\infty} xe^{-x^2} = 0$. Therefore, $f'(0) = 0$.

(b) The Maclaurin series for f is the constant function 0.

(c) The radius of convergence of the series in part (b) is $R = \infty$.

(d) The series in part (b) converges to $f(x)$ only when $x = 0$.

§11.8 Chapter Summary

1. $\lim\limits_{k \to \infty} a_k = \infty.$

2. $\lim\limits_{k \to \infty} a_k = 0.$

3. $\lim\limits_{k \to \infty} a_k = \infty.$

4. $\lim\limits_{k \to \infty} a_k = \infty.$

5. $\lim\limits_{n \to \infty} a_n = \lim\limits_{n \to \infty} \dfrac{n+2}{n^3+4} = \lim\limits_{n \to \infty} \dfrac{1}{3n^2} = 0.$

6. $\lim\limits_{k \to \infty} a_k = \lim\limits_{k \to \infty} \dfrac{\ln(1+k^2)}{\ln(4+k)} = \lim\limits_{k \to \infty} \dfrac{2k(4+k)}{1+k^2} = 2.$

7. $\lim\limits_{k \to \infty} a_k = 0.$

8. $\lim\limits_{k \to \infty} a_k = 0.$

9. Converges absolutely — $\displaystyle\sum_{k=0}^{\infty} \left(\dfrac{1}{k!}\right)^2 \le \sum_{k=0}^{\infty} \dfrac{1}{k!} = e.$

10. Converges absolutely. $\displaystyle\sum_{n=3}^{\infty} \dfrac{1}{n(\ln n)^3} \le \dfrac{1}{3(\ln 3)^3} + \int_{3}^{\infty} \dfrac{dx}{x(\ln x)^3} = \dfrac{1}{3(\ln 3)^3} + \dfrac{1}{2(\ln 3)^2}.$

11. Diverges by the nth term test: $\displaystyle\lim_{n \to \infty} \sum_{k=1}^{n} k^{-1} = \infty.$

12. Diverges. The partial sum of the first $2n$ terms of the series is one-half the partial sum of the first n terms of the harmonic series.

13. Converges—by the integral test: $\displaystyle\sum_{j=1}^{\infty} j \cdot 5^{-j} \le \dfrac{1}{5} + \int_{1}^{\infty} x \cdot 5^{-x}\, dx = \dfrac{1}{5} + \dfrac{1+\ln 5}{5(\ln 5)^2}.$

    NOTE: This series can also be shown to converge via the ratio test:

    $$\lim_{j \to \infty} \dfrac{\frac{j+1}{5^{j+1}}}{\frac{j}{5^j}} = \lim_{j \to \infty} \dfrac{j+1}{j} \cdot \dfrac{5^j}{5^{j+1}} = \lim_{j \to \infty} \dfrac{j+1}{5j} = \dfrac{1}{5} < 1.$$

    Furthermore, since $a_{j+1}/a_j = (j+1)/5j \le 2/5$ when $j \ge 1$, $\displaystyle\sum_{j=1}^{\infty} j \cdot 5^{-j} \le \sum_{j=1}^{\infty} \dfrac{2^{j-1}}{5^j} = \dfrac{1}{3}.$

14. Converges—by the comparison test: $\displaystyle\sum_{j=1}^{\infty} \dfrac{j}{j^4+j-1} < \sum_{j=1}^{\infty} \dfrac{j}{j^4} = \sum_{j=1}^{\infty} \dfrac{1}{j^3} \le \dfrac{3}{2}.$

15. Converges—by the integral test: $\int_0^\infty e^{-x^2}\, dx = \sqrt{\pi}/2$.

$$\sum_{m=0}^\infty e^{-m^2} \le 1 + \int_0^\infty e^{-x^2}\, dx = 1 + \sqrt{\pi}/2.$$

16. Diverges—by the comparison test:

$$\sum_{m=1}^\infty \frac{m^3}{m^4 - 7} = -\frac{1}{6} + \sum_{m=2}^\infty \frac{m^3}{m^4 - 7} > -\frac{1}{6} + \sum_{m=2}^\infty \frac{m^3}{m^4} = -\frac{1}{6} + \sum_{m=2}^\infty \frac{1}{m}.$$

17. Diverges—by the comparison test: $\displaystyle\sum_{k=1}^\infty \frac{k!}{(k+1)! - 1} > \sum_{k=1}^\infty \frac{k!}{(k+1)!} = \sum_{k=1}^\infty \frac{1}{k+1} = \sum_{k=2}^\infty \frac{1}{k}$.

18. Converges—by the integral test: $\displaystyle\int_2^\infty \frac{\ln x}{x^2}\, dx = \int_{\ln 2}^\infty u e^{-u}\, du = \frac{1 + \ln 2}{2}$. Thus,

$$\sum_{j=2}^\infty \frac{\ln j}{j^2} \le \frac{\ln 2}{4} + \frac{1 + \ln 2}{2} = \frac{2 + 3\ln 2}{4}.$$

19. Converges—by the comparison test: $\displaystyle\sum_{k=1}^\infty \frac{\sqrt{k}}{k^2 + 1} < \sum_{k=1}^\infty \frac{1}{k^{3/2}} \le 1 + \int_1^\infty x^{-3/2}\, dx = 3$.

20. The series converges absolutely by the comparison test using $b_k = (2/7)^k$.

$$\left| S - \sum_{k=0}^N a_k \right| \le 0.005 \text{ when } N \ge 4 \text{ since } 2^5/(7^5 + 5) < 0.005. \text{ Using } N = 4,$$
$$S \approx \frac{68917177}{84877260} \approx 0.81196.$$

21. The series converges absolutely by the comparison test: $\displaystyle\sum_{k=0}^\infty \frac{1}{(k+1)\, 2^k} < \sum_{k=0}^\infty \frac{1}{2^k} = 2$.

$$\left| S - \sum_{k=0}^N a_k \right| \le 0.005 \text{ when } N \ge 5 \text{ since } 1/(7 \cdot 2^6) = 1/448 < 0.005. \text{ Using } N = 5,$$
$$S \approx \frac{259}{320} = 0.809375.$$

22. converges absolutely—$\displaystyle\sum_{m=8}^\infty \left| \frac{\sin m}{m^3} \right| < \sum_{m=8}^\infty \frac{1}{m^3}$. Since

$$\left| \sum_{m=8}^\infty \frac{\sin m}{m^3} \right| \le \sum_{m=8}^\infty \frac{1}{m^3} \le \frac{1}{8^3} + \int_8^\infty \frac{dx}{x^3} = \frac{5}{512}, \text{ it follows that } -\frac{5}{512} \le \sum_{m=8}^\infty \frac{\sin m}{m^3} \le \frac{5}{512}.$$

23. converges conditionally—$\displaystyle\sum_{n=1}^\infty \frac{\cos(n\pi)}{n} = \sum_{n=1}^\infty \frac{(-1)^n}{n}$ which is (almost) the alternating harmonic series.

$$-1 < \sum_{n=1}^{\infty} \frac{\cos(n\pi)}{n} = -\ln 2 < -\frac{1}{2}.$$

24. Diverges — $\lim_{k \to \infty} \dfrac{3^k}{k^3 + 3^k} \neq 0.$

25. Diverges by the nth term test. (Since $0 < \pi - e$, $\lim_{k \to \infty} k^{\pi - e} = \infty$.)

26. $\displaystyle\sum_{m=2}^{\infty} \frac{1}{(\ln 3)^m} = \frac{1}{(\ln 3)^2} \frac{1}{1 - (1/\ln 3)} = \frac{1}{(\ln 3)(\ln 3 - 1)}.$

27.

$$\sum_{j=0}^{\infty} \left(\frac{1}{2^j} + \frac{1}{3^j}\right)^2 = \sum_{j=0}^{\infty} \left(\frac{1}{2^{2j}} + \frac{2}{6^j} + \frac{1}{3^{2j}}\right) = \sum_{j=0}^{\infty} \left(\frac{1}{4^j} + \frac{2}{6^j} + \frac{1}{9^j}\right) = \frac{4}{3} + \frac{12}{5} + \frac{9}{8} = \frac{583}{120}.$$

28. $\displaystyle\sum_{k=1}^{\infty} \left(\int_k^{k+1} \frac{dx}{x^2}\right) = \sum_{k=1}^{\infty} \frac{1}{k(k+1)} = \lim_{n \to \infty} \left(1 - \frac{1}{n+1}\right) = 1.$

29. Diverges by the nth term test: $\displaystyle\lim_{m \to \infty} \int_0^m e^{-x^2}\, dx = \sqrt{\pi}/2 \neq 0.$

30. Diverges by the nth term test: $\displaystyle\lim_{n \to \infty} \frac{\ln n}{\ln(3 + n^2)} = \frac{1}{2} \neq 0.$

31. **No.** Using the series representation of $\sin x$ and the alternating series test,
$\sin(1/n) > 1/n - 1/6n^3 = (6n^2 - 1)/6n^3 \geq 5/6n$ for all $n \geq 1$. Thus,
$$\sum_{n=1}^{\infty} \sin(1/n) > \frac{5}{6} \sum_{n=1}^{\infty} \frac{1}{n} = \infty.$$

32. **Yes.** Since $0 < \sin(1/n) < 1/n$ for all $n \geq 1$, $0 < \displaystyle\sum_{n=1}^{\infty} \frac{1}{n} \sin(1/n) < \sum_{n=1}^{\infty} \frac{1}{n^2}$. Thus, the given

series converges by the comparison test.

33. **No.** Since $\lim_{n \to \infty} e^{-1/n} = 1 \neq 0$, the series diverges by the n-th term test.

34. **No.** Using the power series representation of e^x and the alternating series theorem,

$$1 - e^{-1/n} > \frac{1}{n} - \frac{1}{2n^2} = \frac{2n - 1}{2n^2} \geq \frac{2n - n}{2n^2} = \frac{1}{2n}$$

for all $n \geq 1$. Therefore, $\displaystyle\sum_{n=1}^{\infty} \left(1 - e^{-1/n}\right) \geq \frac{1}{2} \sum_{n=1}^{\infty} \frac{1}{n}.$

35. (a) $\ln(n!) = \ln n + \ln(n-1) + \ln(n-2) + \cdots + \ln 2 + \ln 1$. Thus, (since $\ln 1 = 0$) $\ln(n!)$ is a right sum approximation so $\int_1^n \ln x \, dx$. Since $\ln x$ is an increasing function on the interval $[1, n]$, $\ln(n!) > \int_1^n \ln x \, dx = n \ln n - n + 1$. Therefore, $n! > e^{n \ln n - n + 1} = n^n e^{1-n}$.

   (b) Part (a) implies that $\dfrac{b^N}{N!} < \left(\dfrac{be}{N}\right)^N \dfrac{1}{e}$. Thus, $\dfrac{b^N}{N!} < \dfrac{1}{2}$ when $\left(\dfrac{be}{N}\right)^N < \dfrac{e}{2}$. Therefore, since $\dfrac{e}{2} > 1$, any $N > be$ satisfies the given inequality.

36. (a) For all $k \geq 1$, $0 < a_k < 1/k \implies \lim\limits_{k \to \infty} a_k = 0$.

   (b) **No**, the series diverges. Since $a_k = \displaystyle\int_k^\infty \dfrac{dx}{2x^2 - 1} \geq \int_k^\infty \dfrac{dx}{2x^2} = \dfrac{1}{2k}$,

   $$\sum_{k=1}^\infty a_k \geq \dfrac{1}{2} \sum_{k=1}^\infty \dfrac{1}{k} = \infty.$$

   (c) $\displaystyle\sum_{k=1}^\infty (-1)^{k+1} a_k$ converges by the alternating series test — the terms of the series alternate in sign and are decreasing in magnitude (i.e., $a_{k+1} < a_k$).

37. $\left|\dfrac{a_{m+1}}{a_m}\right| = \dfrac{m^2 + 1}{(m+1)^2 + 1} \cdot |x| < 1 \implies R = 1$. The interval of convergence is $[-1, 1]$.

38. $\left|\dfrac{a_{n+1}}{a_n}\right| = \left(\dfrac{n+1}{n}\right)^n \cdot (n+1) \cdot |x| < 1 \implies R = 0$. The interval of convergence is just $x = 0$.

39. The interval of convergence is $(-1/3, 1/3)$.

40. The interval of convergence is $(-\infty, \infty)$.

41. The interval of convergence is $[-1/3, 1/3)$.

42. The interval of convergence is $[-1/3, 1/3]$.

43. $(1, 5)$.

44. $(-\infty, \infty)$.

45. $[-3, 5)$.

46. $[0, 2)$.

47. $[-6, -4]$.

48. $[1/2, 3/2)$.

49. **Cannot** be true. The interval of convergence of a power series is symmetric around and includes its base point ($b = 1$ in this case).

50. **May** be true. (The statement is true when $a_k = 1/k!$ but it is false when $a_k = 1$.)

51. **Must** be true. If the radius of convergence of the power series is 3, then the interval of convergence includes all values of x such that $|x - 1| < 3$.

52. **Cannot** be true. The interval of convergence of this power series must be symmetric about the point $b = 1$.

53. **Cannot** be true. The interval of convergence of the power series is the solution set of the inequality $|x - 1| < 8$. Thus, the radius of convergence of the power series is 8.

54. (a) $\displaystyle\lim_{x \to 1^-} \sum_{k=0}^{\infty} (-1)^k x^k = \lim_{x \to 1^-} \frac{1}{1+x} = \frac{1}{2}.$

    (b) An infinite series converges only if the sequence defined by its partial sums converges. Since the partial sums $\displaystyle\sum_{k=0}^{N} (-1)^k$ are alternately 1 and 0, the infinite series does not converge.

55. $\dfrac{1 - \cos x}{x} = x^{-1}\left(1 - \displaystyle\sum_{k=0}^{\infty}(-1)^k \frac{x^{2k}}{(2k)!}\right) = \displaystyle\sum_{k=1}^{\infty}(-1)^{k+1}\frac{x^{2k-1}}{(2k)!} \implies \lim_{x \to 0} \frac{1 - \cos x}{x} = 0.$

56. $\dfrac{e^x - e^{-x}}{x} = 2\displaystyle\sum_{k=0}^{\infty} \frac{x^{2k}}{(2k+1)!} \implies \lim_{x \to 0} \frac{e^x - e^{-x}}{x} = 2.$

57. $\dfrac{x - \arctan x}{x^3} = \displaystyle\sum_{k=0}^{\infty}(-1)^k \frac{x^{2k}}{2k+3} \implies \lim_{x \to 0} \frac{x - \arctan x}{x^3} = \frac{1}{3}.$

58. $\displaystyle\lim_{x \to 1} \frac{\ln x}{x - 1} = \lim_{w \to 0} \frac{\ln(1+w)}{w} = \lim_{w \to 0} \left(\sum_{k=0}^{\infty}(-1)^k \frac{w^k}{k+1}\right) = 1.$

59. $2^x = e^{x \ln 2} = \displaystyle\sum_{k=0}^{\infty} \frac{(x \ln 2)^k}{k!}; \ R = \infty.$

60. $\cos^2 x = \frac{1}{2}\left(1 + \cos(2x)\right) = \dfrac{1}{2}\left(1 + \displaystyle\sum_{k=0}^{\infty}(-1)^k \frac{(2x)^{2k}}{(2k)!}\right) = \frac{1}{2} + \displaystyle\sum_{k=0}^{\infty}(-1)^k \frac{2^{2k-1}x^{2k}}{(2k)!}; \ R = \infty.$

61. $\dfrac{5 + x}{x^2 + x - 2} = \dfrac{2}{x - 1} - \dfrac{1}{x + 2} = -\dfrac{2}{1 - x} - \dfrac{1}{2}\dfrac{1}{1 + (x/2)}$

    $= -2\displaystyle\sum_{k=0}^{\infty} x^k - \frac{1}{2}\displaystyle\sum_{k=0}^{\infty}(-1)^k \left(\frac{x}{2}\right)^k$

    $= -\displaystyle\sum_{k=0}^{\infty}\left(\frac{2^{k+2} + (-1)^k}{2^{k+1}}\right) x^k; \quad R = 1.$

62. $f(x) = \sin^3(x) = \frac{1}{4}\big(3\sin x - \sin(3x)\big) = \frac{3}{4}\sum_{k=0}^{\infty}(-1)^k \frac{x^{2k+1}}{(2k+1)!} - \frac{1}{4}\sum_{k=0}^{\infty}(-1)^k \frac{(3x)^{2k+1}}{(2k+1)!} =$

$\sum_{k=1}^{\infty}(-1)^{k+1}\frac{3^{2k+1}-3}{4\cdot(2k+1)!}x^{2k+1}; \; R = \infty.$

63. $\dfrac{1}{1-x} = -\dfrac{1}{1+(x-2)} = -\sum_{k=0}^{\infty}(-1)^k(x-2)^k \implies a_k = (-1)^{k+1}.$

64. Since $f'(0) > 0$, the coefficient of x in the Maclaurin series representation of f must be positive; the coefficient of x in the series given is negative.

65. (a) **No.** The Maclaurin series representation of f is

$f(x) = f(0) + f'(0)x + \dfrac{f''(0)}{2}x^2 + \cdots.$ Since f is concave down at $x = 0$, the coefficient of x^2 in the Maclaurin series representation of f is negative.

(b) **Yes.** $g''(0) = \dfrac{3}{4}\big(f'(0)\big)^2\big(g(0)\big)^5 - \dfrac{1}{2}f''(0)\big(g(0)\big)^3 > 0.$

66. (a) The following inequalities are apparent from the figures illustrating the integral test:

$$\int_1^{n+1} a(x)\,dx < \sum_{k=1}^{n} a_k < a_1 + \int_1^{n} a(x)\,dx.$$

Taking $a(x) = 1/x$, these inequalities imply that $\ln(n+1) < H_n < 1 + \ln n$.

(b) First, observe that

$a_n - a_{n+1} = \big(H_n - \ln n\big) - \big(H_{n+1} - \ln(n+1)\big) = \ln(n+1) - \ln n - \dfrac{1}{n+1}.$ Then,

note that $\int_n^{n+1} x^{-1}\,dx = \ln(n+1) - \ln n > \dfrac{1}{n+1}$ since x^{-1} is a decreasing function on the interval $[n, n+1]$. Thus, $a_n > a_{n+1}$.

(c) The sequence a_n is decreasing and bounded below by 0 (since $H_n - \ln n > \ln(n+1) - \ln n > 0$). Thus, it converges.

(d) $\int_x^{\infty} f'(t)\,dt = \lim_{a\to\infty}\int_x^{a} f'(t)\,dt = \lim_{a\to\infty}\big(f(a) - f(x)\big) = -f(x).$

[NOTE: $f(a) = \ln\left(\dfrac{a+1}{a}\right) - \dfrac{1}{a+1} = \ln\left(1+\dfrac{1}{a}\right) - \dfrac{1}{a+1}.$]

(e) When $x > 0$, $f(x) = -\int_x^{\infty} f'(t)\,dt > \int_x^{\infty}\dfrac{dt}{(t+1)^3} = \dfrac{1}{2(x+1)^2}.$

(f) Let

$$S_N = \sum_{k=n}^{N}(a_k - a_{k+1})$$

$$= (a_n - a_{n+1}) + (a_{n+1} - a_{n+2}) + \cdots + (a_N - a_{N+1}) = a_n - a_{N+1}.$$

Since $\gamma = \lim_{n \to \infty} a_n$, $\sum_{k=n}^{\infty} (a_k - a_{k+1}) = \lim_{N \to \infty} S_N = a_n - \lim_{N \to \infty} a_{N+1} = a_n - \gamma$.

(g) Since f is a decreasing function, the integral test implies that $\int_n^{\infty} f(x)\,dx \le \sum_{k=n}^{\infty} f(k)$.

Therefore, part (b) implies that $\int_n^{\infty} f(x)\,dx > \dfrac{1}{2(n+1)}$.

To get the upper bound on $a_n - \gamma$, note that $f(k) < \dfrac{1}{2}\left(\dfrac{1}{k} - \dfrac{1}{k+1}\right)$. (Apply the

trapezoid rule to $\int_k^{k+1} dx/x$.) This inequality implies that

$$\sum_{k=n}^{\infty} f(k) < \sum_{k=n}^{\infty} \dfrac{1}{2}\left(\dfrac{1}{k} - \dfrac{1}{k+1}\right) = \dfrac{1}{2n}.$$

67. Since $\lim_{n \to \infty} \left|\dfrac{a_{n+1}}{a_n}\right| = \lim_{n \to \infty} \dfrac{|r-n|\cdot|x|}{n+1} < 1$ if $|x| < 1$, the binomial series converges if $|x| < 1$.

68. Let $r = 1/2$.

69. The binomial series for $(1+u)^3$ terminates after a finite number of terms since $r = 3$ is an integer: $(1+u)^3 = 1 + 3u + 3u^2 + u^3$. Therefore,
$f(x) = (1 + x^4)^3 = 1 + 3x^4 + 3x^8 + x^{12}$.

70. $g(x) = \sqrt[3]{1 - x^2} \approx 1 - x^2/3 - x^4/9 - 5x^6/81$.

71. $g(x) = (1 + x^2)^{-3/2} \approx 1 - 3x^2/2 + 15x^4/8 - 35x^6/16$.

72. $g(x) = \arcsin x = \displaystyle\int \dfrac{dx}{\sqrt{1 - x^2}}$

$= \displaystyle\int \left(1 + \dfrac{x^2}{2} + \dfrac{3x^4}{8} + \dfrac{5x^6}{16} + \cdots\right) dx \approx x + \dfrac{x^3}{6} + \dfrac{3x^5}{40} + \dfrac{5x^7}{112}$.

73. Since $\sqrt{1 + u} = 1 + \dfrac{u}{2} - \dfrac{u^2}{8} + \dfrac{u^3}{16} - \dfrac{5u^4}{128} \pm \cdots$,

$\sqrt{1 + x^3} = 1 + \dfrac{x^3}{2} - \dfrac{x^6}{8} + \dfrac{x^9}{16} - \dfrac{5x^{12}}{128} \pm \cdots$. Thus,

$\displaystyle\int_0^{2/5} \sqrt{1 + x^3}\,dx = \int_0^{2/5} \left(1 + \dfrac{x^3}{2} - \dfrac{x^6}{8} + \dfrac{x^9}{16} - \dfrac{5x^{12}}{128} \pm \cdots\right) dx$

$= \dfrac{2}{5} + \dfrac{1}{8}\left(\dfrac{2}{5}\right)^4 - \dfrac{1}{56}\left(\dfrac{2}{5}\right)^7 + \dfrac{1}{160}\left(\dfrac{2}{5}\right)^{10} - \dfrac{5}{1664}\left(\dfrac{2}{5}\right)^{13} \pm \cdots$.

Now, since this is an alternating series (after the first term) and $2^7/(56 \cdot 5^7) < 5 \times 10^{-4}$, the value of the integral is approximated to the desired accuracy by
$2/5 + 2^4/(8 \cdot 5^4) = 252/625 = 0.4032$.

74. (a) $f'(x) = \displaystyle\sum_{n=1}^{\infty} \frac{r(r-1)(r-2)\cdots(r-n+1)}{(n-1)!} x^{(n-1)}$. The coefficient of x^n in the series for $(1+x)f'(x)$ is

$$\frac{r(r-1)(r-2)\cdots(r-n+1)(r-n)}{n!} + \frac{r(r-1)(r-2)\cdots(r-n+1)}{(n-1)!}$$

$$= \frac{(r-n)\cdot r(r-1)(r-2)\cdots(r-n+1)}{n!} + \frac{n\cdot r(r-1)(r-2)\cdots(r-n+1)}{n!}$$

$$= \frac{r\cdot r(r-1)(r-2)\cdots(r-n+1)}{n!}.$$

Since this is also the coefficient of x^n in the series for $rf(x)$, the result follows.

(b) $g'(x) = \dfrac{f'(x)}{(1+x)^r} - \dfrac{rf(x)}{(1+x)(1+x)^r} = \dfrac{f'(x)}{(1+x)^r} - \dfrac{(1+x)f'(x)}{(1+x)(1+x)^r} = 0.$

(c) The result in part (c) implies that g is a constant function. Since $g(0) = 1$, $g(x) = 1$ and so $f(x) = (1+x)^r$.

75. (a) By the binomial power series,

$$f(x) = (1+x)^r = 1 + \sum_{n=1}^{\infty} \frac{r(r-1)(r-2)\cdots(r-n+1)}{n!} x^n.$$

Thus, $g(x) = (1 - x^2)^{-1/2}$ can be written as

$$1 + \sum_{n=1}^{\infty} \frac{(-1/2)(-3/2)(-5/2)\cdots(1/2-n)}{n!}(-x^2)^n$$

$$= 1 + \sum_{n=1}^{\infty} \frac{(-1)^n(-1)(-3)(-5)\cdots(1-2n)}{2^n \cdot n!} x^{2n}$$

$$= 1 + \sum_{n=1}^{\infty} \frac{1\cdot 3\cdot 5\cdots(2n-1)}{2\cdot 4\cdot 6\cdots(2n)} x^{2n}$$

To finish, we integrate:

$$\int \left(1 + \sum_{n=1}^{\infty} \frac{1\cdot 3\cdot 5\cdots(2n-1)}{2\cdot 4\cdot 6\cdots(2n)} x^{2n}\right) dx = x + \sum_{n=1}^{\infty} \frac{1\cdot 3\cdot 5\cdots(2n-1)}{2\cdot 4\cdot 6\cdots(2n)} \frac{x^{2n+1}}{2n+1}$$

$$= \int_0^x \frac{dt}{\sqrt{1-t^2}}$$

$$= \arcsin x.$$

(b) Let $u = \arcsin x$. Then $du = \frac{dx}{\sqrt{1-x^2}}$ and $\sin u = x$. Thus,

$$\int_0^1 \frac{x^{2n+1}}{\sqrt{1-x^2}}\,dx = \int_0^{\pi/2} \sin^{2n+1} u\,du$$

$$= \frac{2\cdot 4\cdot 6\cdots(2n)}{3\cdot 5\cdot 7\cdots(2n+1)}.$$

(c) $\displaystyle\int_0^1 \frac{\arcsin x}{\sqrt{1-x^2}}\, dx = \int_0^1 \frac{\arcsin x}{x^{2n+1}} \cdot \frac{x^{2n+1}}{\sqrt{1-x^2}}\, dx$

$\displaystyle = \int_0^1 \left(x^{-2n} + \sum_{n=1}^{\infty} \frac{1 \cdot 3 \cdot 5 \cdots (2n-1)}{2 \cdot 4 \cdot 6 \cdots (2n)} \frac{1}{2n+1} \right) \frac{x^{2n+1}}{\sqrt{1-x^2}}\, dx$

$\displaystyle = \int_0^1 \frac{x}{\sqrt{1-x^2}}\, dx + \left(\sum_{n=1}^{\infty} \frac{1 \cdot 3 \cdot 5 \cdots (2n-1)}{2 \cdot 4 \cdot 6 \cdots (2n)} \frac{1}{2n+1} \right) \int_0^1 \frac{x^{2n+1}}{\sqrt{1-x^2}}\, dx$

$\displaystyle = 1 + \sum_{k=1}^{\infty} \frac{1}{(2k+1)^2} = \sum_{k=0}^{\infty} \frac{1}{(2k+1)^2}.$

(d) Let $u = \arcsin x$. Then $du = 1/\sqrt{1-x^2}\,dx$. Thus,

$$\int_0^1 \frac{\arcsin x}{\sqrt{1-x^2}}\, dx = \int_0^{\pi/2} u\, du = \frac{\pi^2}{8}$$

(e) After writing out a few terms, it is clear that

$$\sum_{k=1}^{\infty} \frac{1}{k^2} = \sum_{k=0}^{\infty} \frac{1}{(2k+1)^2} + \sum_{k=1}^{\infty} \frac{1}{(2k)^2}.$$

Since the summation on the left hand side is a convergent *p*-series, we can assign to it some limit L. Then, the above equation can be rewritten as $L = \pi^2/8 + L/4$. Solving for L yields $L = \pi^2/6$. Thus, $\displaystyle\sum_{k=1}^{\infty} \frac{1}{k^2} = \frac{\pi^2}{6}.$

76. (a) By part (c) of the Exercise 19 in Section 9.2, $R_n = \dfrac{1}{n!} \displaystyle\int_0^x (x-t)^n f^{(n+1)}(t)\, dt$. Let
$f(t) = (1+t)^r$. After differentiating f a few times, it is easy to see that
$f'(t) = r(r-1)(r-2)\cdots(r-n)(1+t)^{r-(n+1)}$. Thus
$R_n = \dfrac{r \cdot (r-1) \cdot (r-2) \cdots (r-n)}{n!} \displaystyle\int_0^x \frac{(x-t)^n}{(1+t)^{n+1-r}}\, dt.$

(b) Since $t \geq 0$, $\dfrac{|x-t|}{1+t} \leq |x-t| \leq |x|$.

(c) By parts (a) and (b),

$$|R_n(x)| \leq \frac{|r \cdot (r-1) \cdot (r-2) \cdots (r-n)|}{n!} x^n \int_0^x |(1+t)^{r-1}|\, dt$$

$$= \frac{|(r-1) \cdot (r-2) \cdots (r-n)|}{n!} x^n |(1+x)^r - 1|.$$

(d) If $0 \leq x < 1$, then
$\displaystyle\lim_{n \to \infty} |R_n(x)| = \lim_{n \to \infty} \frac{|(r-1) \cdot (r-2) \cdots (r-n)|}{n!} x^n |(1+x)^r - 1| = 0$. Thus, the
binomial series converges to f along the interval $[0, 1)$.

(e) If $-1 < x \le t < 0$, then $\dfrac{-|x-t|}{1+t} \le |x|$. Thus,

$$|R_n(x)| \le \frac{|r \cdot (r-1) \cdot (r-2) \cdots (r-n)|}{n!}|x|^n \int_0^x |(1+t)^{r-1}|\, dt$$

$$= \frac{|r \cdot (r-1) \cdot (r-2) \cdots (r-n)|}{n!}|x|^n|1-(1+x)^r|.$$

If $-1 < x < 0$, this series converges converges as $n \to \infty$, since $\lim_{n\to\infty} |x|^n = 0$.

77. (a) Notice that $\displaystyle\lim_{m\to\infty} \sum_{k=0}^m \frac{(-1)^k}{k!} = e^{-1}$. Furthermore, since $\displaystyle\sum_{k=0}^m \frac{(-1)^k}{k!}$ is an alternating

series, $\left| \dfrac{1}{e} - \displaystyle\sum_{k=0}^m \dfrac{(-1)^k}{k!} \right| \le \dfrac{1}{(m+1)!}$. Therefore,

$$m!\left| \frac{1}{e} - \sum_{k=0}^m \frac{(-1)^k}{k!} \right| \le \frac{m!}{(m+1)!} = \frac{1}{m+1}.$$

(b) $m!/e = (n \cdot m!)/m = n \cdot (m-1)!$. Since this final expression is a product of integers, $m!/e$ is also an integer.

(c) Let $a_k = m!/k!$, where k is an integer such that $0 \le k \le m$. Then, $a_k = m(m-1)(m-2) \cdots (m-(m-k-1))$. Since a_k is a product of integers, it, too, is an integer. Thus, since $m! \displaystyle\sum_{k=0}^m \dfrac{(-1)^k}{k!}$ is the sum of alternatingly positive and negative integers, the summation itself is an integer.

(d) By assumption, m is a positive integer. Thus, $1/(m+1) \le 1/2$ for all m. Since N is the product of two non-negative integer values, it follows that $N = 0$.

(e) By the previous part $m!\left| \dfrac{1}{e} - \displaystyle\sum_{k=0}^m \dfrac{(-1)^k}{k!} \right| = 0$. Since $m! > 0$, it follows that

$$\frac{1}{e} = \sum_{k=0}^m \frac{(-1)^k}{k!} = \sum_{k=0}^\infty \frac{(-1)^k}{k!} - \sum_{k=m+1}^\infty \frac{(-1)^k}{k!}.$$

Thus, $\displaystyle\sum_{k=m+1}^\infty \dfrac{(-1)^k}{k!} = 0$. However, this is false, since the terms of the summation are both alternating and decreasing. Therefore, e is irrational.

78. (a) The sum of k numbers is less than or equal to k times the largest summand; similarly, it is greater than or equal to k times the smallest summand. Thus, since the sequence $\{a_n\}$ is decreasing,

$$2^{m-1}a_{2^m} \le a_{2^{m-1}+1} + a_{2^{m-1}+2} + \cdots + a_{2^m} \le 2^{m-1}a_{2^{m-1}+1} \le 2^{m-1}a_{2^{m-1}}.$$

(b) $\displaystyle\sum_{k=2}^{2^m} a_k = \sum_{k=1}^{m}\left(\sum_{j=2^{k-1}+1}^{2^k} a_j\right) \geq \sum_{k=1}^{m} 2^{k-1}a_{2^k} = \frac{1}{2}\sum_{k=1}^{m} 2^k a_{2^k}.$

(c) Since the terms of the series are positive, the sequence of partial sums is increasing. Therefore, if $\displaystyle\frac{1}{2}\sum_{k=1}^{\infty} 2^k a_{2^k}$ diverges, the sequence of its partial sums is unbounded. Since

the partial sums of this series are lower bounds for the partial sums of the series $\displaystyle\sum_{k=1}^{\infty} a_k$,

the latter series must also diverge.

(d) Since the terms of the series are positive,

$$\sum_{k=2}^{n} a_k \leq \sum_{k=2}^{2^m} a_k = \sum_{k=1}^{m}\left(\sum_{j=2^{k-1}+1}^{2^k} a_j\right) \leq \sum_{k=1}^{m} 2^{k-1}a_{2^{k-1}}.$$

(e) Suppose that $\displaystyle\sum_{k=0}^{\infty} 2^k a_{2^k}$ converges. Part (d) implies that the partial sums of the series

$\displaystyle\sum_{k=2}^{\infty} a_k$ are bounded above by $\displaystyle\sum_{k=0}^{\infty} 2^k a_{2^k}$. Furthermore, since $a_k > 0$ for all $k \geq 1$, the

partial sums of the series $\displaystyle\sum_{k=2}^{\infty} a_k$ form an increasing sequence. Together, these results

imply that the series $\displaystyle\sum_{k=2}^{\infty} a_k$ converges (the sequence of its partial sums is increasing and

bounded above).

79. Let $a_k = 1/k$. Then $\displaystyle\sum_{k=1}^{\infty} 2^k a_{2^k} = \sum_{k=1}^{\infty} 1$ which is obviously a divergent series. Therefore,

part (c) of the previous exercise implies that $\displaystyle\sum_{k=1}^{\infty} a_k$ diverges.

80. Let $a_k = 1/k^p$. According to Exercise 78, the series $\displaystyle\sum_{k=1}^{\infty} a_k$ converges if and only if the series

$$\sum_{k=0}^{\infty} 2^k a_{2^k} = \sum_{k=0}^{\infty} \frac{2^k}{2^{kp}} = \sum_{k=0}^{\infty}\left(\frac{1}{2^{p-1}}\right)^k = \sum_{k=0}^{\infty}\left(2^{1-p}\right)^k$$

converges. Since the series on the right is a geometric series, it converges only if $2^{1-p} < 1$.

81. Let $a_k = \frac{1}{k(\ln k)^p}$. The series $\sum_{k=2}^{\infty} a_k$ converges if and only if the series

$$\sum_{k=1}^{\infty} 2^k a_{2^k} = \sum_{k=1}^{\infty} \frac{2^k}{2^k \left(\ln 2^k\right)^p} = \sum_{k=1}^{\infty} \frac{1}{k^p (\ln 2)^p} = \frac{1}{(\ln 2)^p} \sum_{k=1}^{\infty} \frac{1}{k^p}$$

converges. By part (a), this series converges only when $p > 1$.

82. $\displaystyle\int_0^{\infty} e^{-t} \sin(xt)\, dt = \int_0^{\infty} e^{-t} \left(\sum_{k=0}^{\infty} (-1)^k \frac{(xt)^{2k+1}}{(2k+1)!} \right) dt$

$$= \sum_{k=0}^{\infty} (-1)^k \frac{x^{2k+1}}{(2k+1)!} \left(\int_0^{\infty} e^{-t} t^{2k+1}\, dt \right)$$

$$= \sum_{k=0}^{\infty} (-1)^k x^{2k+1} = \frac{x}{1+x^2} \quad \text{when } |x| < 1.$$

83. (a) Using the trigonometric identity given,

$$\frac{1}{2^k} \tan\left(\frac{x}{2^k}\right) = \frac{1}{2^k} \cot\left(\frac{x}{2^k}\right) - \frac{1}{2^{k-1}} \cot\left(\frac{x}{2^{k-1}}\right).$$

Thus, $\displaystyle\sum_{k=1}^{n} \frac{1}{2^k} \tan\left(\frac{x}{2^k}\right)$ can be written as a telescoping sum and

$$\sum_{k=1}^{n} \frac{1}{2^k} \tan\left(\frac{x}{2^k}\right) = \frac{1}{2^n} \cot\left(\frac{x}{2^n}\right) - \cot x.$$

(b) Using l'Hôpital's Rule,

$$\lim_{n \to \infty} \sum_{k=1}^{n} \frac{1}{2^k} \tan\left(\frac{x}{2^k}\right) = \lim_{n \to \infty} \frac{1}{2^n} \cot\left(\frac{x}{2^n}\right) - \cot x = \frac{1}{x} - \cot x.$$

84. (a) Since $0 \le \tan x \le 1$ when $0 \le x \le \pi/4$,
$$I_{n+1} = \int_0^{\pi/4} \tan^{n+1} x\, dx \le \int_0^{\pi/4} \tan^n x\, dx = I_n \text{ for any integer } n \ge 0.$$

(b) $\displaystyle I_n + I_{n-2} = \int_0^{\pi/4} \tan^n x\, dx + \int_0^{\pi/4} \tan^{n-2} x\, dx$

$$= \int_0^{\pi/4} (1 + \tan^2 x) \tan^{n-2} x\, dx = \int_0^{\pi/4} \sec^2 x \tan^{n-2} x\, dx$$

$$= \int_0^1 u^{n-2}\, du = \frac{1}{n-1}.$$

(c) Since $\{ I_n \}$ is a nonincreasing sequence, $\dfrac{1}{n-1} = I_n + I_{n-2} \geq 2I_n$ so $\dfrac{1}{2(n-1)} \geq I_n$.

Similarly, $\dfrac{1}{n+1} = I_{n+2} + I_n \leq 2I_n$ so $\dfrac{1}{2(n+1)} \leq I_n$.

(d) Part (b) may be rewritten $I_n = \dfrac{1}{n-1} - I_{n-2} = \dfrac{1}{n-1} - \displaystyle\int_0^{\pi/4} \tan^{n-2} x \, dx$.

(e) First, note that $\displaystyle\int_0^{\pi/4} \tan^0 x \, dx = \dfrac{\pi}{4}$. Then, part (d) can be used repeatedly to show that

$$I_{2n} = \frac{1}{2n-1} - I_{2n-2} = \frac{1}{2n-1} - \left(\frac{1}{2n-3} - I_{2n-4} \right)$$

$$= \frac{1}{2n-1} - \frac{1}{2n-3} + \left(\frac{1}{2n-5} - I_{2n-6} \right) = \cdots = (-1)^n \left(\frac{\pi}{4} + \sum_{k=1}^{n} \frac{(-1)^k}{2k-1} \right)$$

for all integers $n \geq 1$.

(f) It follows from part (c) that $\displaystyle\lim_{n\to\infty} I_n = 0$. Therefore,

$$-\frac{\pi}{4} = \lim_{n\to\infty} \sum_{k=1}^{n} \frac{(-1)^k}{2k-1} = \lim_{n\to\infty} \sum_{k=0}^{n-1} \frac{(-1)^{k+1}}{2k+1}$$

$$= \sum_{k=0}^{\infty} \frac{(-1)^{k+1}}{2k+1} = -\sum_{k=0}^{\infty} \frac{(-1)^k}{2k+1}.$$

(g) First, note that $\displaystyle\int_0^{\pi/4} \tan^1 x \, dx = \tfrac{1}{2}\ln 2$. Then, part (d) can be used repeatedly to show that

$$I_{2n+1} = \frac{1}{2n} - I_{2n-1} = \frac{1}{2n} - \left(\frac{1}{2n-2} - I_{2n-3} \right) = \frac{1}{2n} - \frac{1}{2n-2} + \left(\frac{1}{2n-4} - I_{2n-5} \right)$$

$$= \cdots = (-1)^n \left(\tfrac{1}{2}\ln 2 + \sum_{k=1}^{n} \frac{(-1)^k}{2k} \right)$$

for all integers $n \geq 1$.

(h) It follows from part (c) that $\displaystyle\lim_{n\to\infty} I_n = 0$. Therefore,

$$-\frac{\ln 2}{2} = \lim_{n\to\infty} \sum_{k=1}^{n} \frac{(-1)^k}{2k} = \lim_{n\to\infty} \frac{1}{2} \sum_{k=1}^{n} \frac{(-1)^k}{k} = \frac{1}{2} \sum_{k=1}^{\infty} \frac{(-1)^k}{k}.$$

V Vectors and Polar Coordinates

§V.1 Vectors and vector-valued functions

1. $u + v = (1, 2) + (2, 3) = (3, 5)$.

2. $u - v = (1, 2) - (2, 3) = (-1, -1)$.

3. $2u - 3v = 2(1, 2) - 3(2, 3) = (-4, -5)$.

4. $u + v/2 + w/3 = (1, 2) + \frac{1}{2}(2, 3) + \frac{1}{3}(-2, 1) = (4/3, 23/6)$.

5. $|u| = \sqrt{1^2 + 2^2} = \sqrt{5}$.

6. $|w| = \sqrt{(-2)^2 + 1^2} = \sqrt{5}$.

7. $|3u + 2w| = |(-1, 8)| = \sqrt{(-1)^2 + 8^2} = \sqrt{65}$.

8. $|2v - 3u| = |(1, 0)| = \sqrt{1^2 + 0^2} = 1$.

9. $u + (v + w) = u + \big((c, d) + (e, f)\big) = u + (c + e, d + f) = (a + c + e, b + d + f)$ and
   $(u + v) + w = (a + c, b + d) + w = (a + c + e, b + d + f)$. Thus,
   $u + (v + w) = (u + v) + w$.

10. $r(v + w) = r(c + e, d + f) = \big(r(c + e), r(d + f)\big)$ and
    $rv + rw = (rc, rc) + (re, rf) = \big(r(c + e), r(d + f)\big)$. Thus, $r(v + w) = rv + rw$.

11. (a) The diagonal vector $v + w$ has components $(3, 4)$.

    (b) The northwest-pointing diagonal vector is $w - v = (-1, 2)$.

    (c) The southeast-pointing diagonal vector is $v - w = (1, -2)$.

12. The shortest distance between two points is a straight line.

21. (a) $L(0) = (1, 2)$; $L(1) = (3, 5)$; $L(2) = (5, 8)$; $L(-1) = (-1, -1)$.

    (b) If the domain of L is restricted to $t \geq 0$, then the image is the ray (or half-line) that
    starts at $P(1, 2)$ and points in the same direction as the vector $(2, 3)$.

    (c) If the domain of L is restricted to $-1 \leq t \leq 1$, then the image is the line segment from
    $L(-1) = (-1, -1)$ to $L(1) = (3, 5)$.

22. (a) $L(t) = (a, b) + t(c, d)$. Therefore, $L(0) = (a, b)$, $L(1) = (a + c, b + d)$,
    $L(2) = (a + 2c, b + 2d)$, and $L(-1) = (a - c, b - d)$.

    (b) ray from (a, b) in the same direction as the vector (c, d).

    (c) line segment from $(a - c, b - d)$ to $(a + c, b + d)$.

23. The particle's velocity vector at time t is $v(t) = \left(-2 + 3t^2, 8t^3\right)$ and the particle's speed at time t is

$$s(t) = |v(t)| = \sqrt{x'(t)^2 + y'(t)^2} = \sqrt{(-2 + 3t^3)^2 + (8t^3)^2}.$$

Therefore, at time $t = 1$, the particle's speed is $\sqrt{65}$.

24. (a) The curve is the straight line joining the origin to the point $(5000, 10000)$.

  (b) Since $p(t) = \frac{t^2}{2}(1, 2)$, we have $x(t) = t^2/2$ and $y(t) = t^2$. Thus $y = 2x$—the equation of the straight line mentioned above.

  (c) For $t \geq 0$, the speed function is $s(t) = |v(t)| = |t(1, 2)| = t\sqrt{5}$. Thus the arc length is $\sqrt{5}\int_0^{100} t\, dt = 5000\sqrt{5} \approx 11180.34$.

25. (a) The curve is the straight line joining the origin to the point $(5000, -5000)$.

  (b) Since $p(t) = \frac{t^2}{2}(1, -1)$, we have $x(t) = t^2/2$ and $y(t) = -t^2/2$. Thus $y = -x$—the equation of the straight line mentioned above.

  (c) For $t \geq 0$, the speed function is $s(t) = |v(t)| = |t(1, -1)| = t\sqrt{2}$. Thus the arc length is $\sqrt{2}\int_0^{100} t\, dt = 5000\sqrt{2} \approx 7071.07$.

26. (a) $v(t) = t(0, -1) + (4, 4)$; $p(t) = \dfrac{t^2}{2}(0, -1) + t(4, 4)$.

  (b) Plotting $x = 4t$ and $y = 4t - t^2/2$ for $0 \leq t \leq 10$ gives a parabola.

  (c) Eliminating t from the equations above gives $y = x - x^2/32$.

  (d) The speed function is $s(t) = |v(t)| = |(4, 4 - t)| = \sqrt{32 - 8t + t^2}$. Thus the arc length is $\int_0^{10} \sqrt{32 - 8t + t^2}\, dt \approx 49.56$. (This is tricky to integrate, but it's easy to estimate numerically, e.g., with the midpoint rule.)

27. (a) The assumption that the projectile travels under free fall conditions implies that $a(t) = (0, -g)$. Since the initial angle is $\pi/3$ and the initial speed is 100 meters per second, $v(t) = \left(100\cos(\pi/3), 100\sin(\pi/3)\right) + t(0, -g) = (50, 50\sqrt{3}) + t(0, -g) = (50, 50\sqrt{3} - gt)$. Finally, since the projectile is at $(0, 0)$ at time $t = 0$, the position function is $p(t) = t(50, 50\sqrt{3}) + t^2(0, -g/2)$.

  (b) The projectile touches down when its y-coordinate is zero: at the time $t_0 > 0$ that is the solution of the equation $50\sqrt{3}t_0 - gt_0^2/2 = 0$. Therefore, $t_0 = 100\sqrt{3}/g \approx 17.67$.

  (d) The projectile is at its maximum height when the velocity vector is horizontal (i.e., when its y-component is zero). Thus, the projectile reaches its maximum height at time $t = 50\sqrt{3}/g$. At this time, the height of the projectile is $50^2 \cdot 3/g - 50^2 \cdot 3/2g = 50^2 \cdot 3/2g \approx 382.65$.

  (e) At time $t = 50\sqrt{3}/g$, the projectile's velocity is $(50, 0)$ and it's speed is 50 meters per second.

28. (a) $v(t) = (s_0\cos\alpha, s_0\sin\alpha) + t(0, -g)$; $p(t) = t(s_0\cos\alpha, s_0\sin\alpha) + t^2(0, -g/2)$.

  (b) $t = (s_0/g)\sin\alpha$.

    (c) speed $= s_0 \cos \alpha$; velocity $= (s_0 \cos \alpha, 0)$.

    (d) When $t = (2s_0/g) \sin \alpha$, speed $= s_0$ and velocity $= (s_0 \cos \alpha, -s_0 \sin \alpha)$.

29.   (a) $\boldsymbol{a}(1) = (13, -1)$.

    (b) $\boldsymbol{p}(0) = (-31/3, 0)$.

    (c) $\displaystyle\int_0^2 \sqrt{25t^4 + 30t^3 - 30t^2 - 26t + 17}\, dt$.

30.   (a) Yes; at $t = 1$ the particle is at $(-2, -1)$.

    (b) Yes; at $t = -1$ the particle is at $(2, 3)$.

    (c) No.

31.   (a) $\boldsymbol{v}(t) = \boldsymbol{v}(0) + (0, -t) = (1, -t)$.

    (b) $\boldsymbol{p}(t) = \boldsymbol{p}(0) + (t, -t^2/2) = (t, -t^2/2)$.

    (c) $\displaystyle\int_0^5 |\boldsymbol{v}(t)|\, dt = \int_0^5 \sqrt{1 + t^2}\, dt$.

32. The arc length integral is $\displaystyle\int_0^1 \sqrt{1^2 + 3^2}\, dt = \int_0^1 \sqrt{10}\, dt = \sqrt{10}$.

33. The arc length integral is
$$\int_0^\pi \sqrt{\left(-2\sin(2t)\right)^2 + \left(2\cos(2t)\right)^2}\, dt = 2\int_0^\pi \sqrt{\sin^2(2t) + \cos^2(2t)}\, dt = 2\int_0^\pi 1\, dt = 2\pi.$$

34. The arc length integral is $\displaystyle\int_0^{2\pi/3} \sqrt{\left(3\cos(3t)\right)^2 + \left(-3\sin(3t)\right)^2}\, dt = 3\int_0^{2\pi/3} 1\, dt = 2\pi$.

35. The arc length integral is $\displaystyle\int_0^{2\pi} \sqrt{(3\cos t)^2 + (-\sin t)^2}\, dt = \int_0^{2\pi} \sqrt{8\cos^2 t + 1}\, dt \approx 13.365$.

36. The arc length integral is

$$\int_0^{4\pi} \sqrt{(\cos t - t\sin t)^2 + (\sin t + t\cos t)^2}\, dt$$

$$= \int_0^{4\pi} \sqrt{t^2 + 1}\, dt = \left. \frac{t\sqrt{t^2+1} + \ln\left|t + \sqrt{t^2+1}\right|}{2} \right]_0^{4\pi} \approx 80.82.$$

37. $\left|\boldsymbol{r}'(t)\right| = \sqrt{1^2 + 3^2} = \sqrt{10}$.

38. $\left|\boldsymbol{r}'(t)\right| = \sqrt{\left(-2\sin(2t)\right)^2 + \left(2\cos(2t)\right)^2} = 2$.

39. $\left|\boldsymbol{r}'(t)\right| = \sqrt{\left(3\cos(3t)\right)^2 + \left(-3\sin(3t)\right)^2} = 3$.

---

40. $|r'(t)| = \sqrt{(3\cos t)^2 + (-\sin t)^2} = \sqrt{8\cos^2 t + 1}.$

41. $|r'(t)| = \sqrt{(\cos t - t\sin t)^2 + (\sin t + t\cos t)^2} = \sqrt{1 + t^2}.$

42. (a) The curve is a spiral around the z-axis; it makes two loops.

   (b) The position, velocity, and acceleration vectors are, respectively, $p(t) = (\sin t, t)$, $v(t) = (\cos t, 1)$, and $a(t) = (-\sin t, 0)$. At $t = 0$, these vectors are $p(0) = (0, 0)$, $v(0) = (1, 1)$, and $a(0) = (0, 0)$.

   (c) At $t = \pi$ we have $p(\pi) = (0, \pi)$ and $v(t) = (-1, 1)$. Therefore, the tangent line ℓ has equation $X(t) = (0, \pi) + t(-1, 1)$.

   (d) Use technology to plot both the curve and the tangent line; method depends on the technology.

   (e) The arc length is $\displaystyle\int_0^{4\pi} \sqrt{1 + \cos^2 t}\, dt \approx 15.281.$

43. In parametric form, the equation of the logarithmic spiral between $\theta = 0$ and $\theta = \pi$ is $x(t) = e^t \cos t,\ y(t) = e^t \sin t,\ 0 \le t \le \pi$. Therefore, the length of the logarithmic spiral $r = e^\theta$ between $\theta = 0$ and $\theta = \pi$ is

$$\int_0^\pi \sqrt{e^{2t}(\cos t - \sin t)^2 + e^{2t}(\sin t + \cos t)^2}\, dt = \int_0^\pi \sqrt{2}\, e^t\, dt = \sqrt{2}\,(e^\pi - 1) \approx 31.312.$$

§V.2 Polar Coordinates and Polar Curves

1. $(\pi, 0)$; $(-\pi, \pi)$; $(\pi, 2\pi)$.

2. $(\pi, \pi/2)$; $(\pi, -3\pi/2)$; $(-\pi, -\pi/2)$; $(-\pi, 3\pi/2)$.

3. $(\sqrt{2}, \pi/4)$; $(\sqrt{2}, -7\pi/4)$; $(-\sqrt{2}, -3\pi/4)$; $(-\sqrt{2}, 5\pi/4)$.

4. $(\sqrt{2}, 3\pi/4)$; $(\sqrt{2}, -5\pi/4)$; $(-\sqrt{2}, 7\pi/4)$; $(-\sqrt{2}, -\pi/4)$.

5. $(\sqrt{5}, 1.107)$; $(\sqrt{5}, -5.176)$;
   $(-\sqrt{5}, 4.249)$.

6. $(\sqrt{5}, 2.034)$; $(\sqrt{5}, -4.249)$; $(-\sqrt{5}, 5.176)$.

7. $(\sqrt{17}, 1.326)$; $(\sqrt{17}, -4.957)$; $(-\sqrt{17}, 4.467)$.

8. $(\sqrt{101}, 0.0997)$; $(\sqrt{101}, 6.3829)$; $(-\sqrt{101}, 3.2413)$.

9. $(\sqrt{2}, \sqrt{2})$.

10. $(\sqrt{2}, \sqrt{2})$.

11. $(\sqrt{3}/2, 1/2)$.

12. $(42, 0)$.

13. $(0.5403, 0.8415)$.

14. $(-\cos 2, -\sin 2) \approx (0.4161, -0.9093)$.

15. $(\sqrt{2}, \sqrt{2})$.

16. $(3\sqrt{5}/5, 6\sqrt{5}/5)$.

17. $x = 2$.

18. $x^2 + y^2 = 16$.

19. $\theta = \pi/3 \implies \tan\theta = \tan(\pi/3) = \sqrt{3} = y/x \implies y = \sqrt{3}x$.

20. $r = 2\sin\theta \implies r^2 = 2r\sin\theta \implies x^2 + y^2 = 2y$.

21. $r = 3$.

22. $r = 4\csc\theta$.

23. $\tan\theta = 2$.

24. $r = 2\cos\theta$.

25. (a) The polar points $(1, 0)$, $(1, 2\pi)$, and $(-1, \pi)$ all represent the Cartesian point $(1, 0)$. (Use $x = r\cos\theta$ and $y = r\sin\theta$.)

    (b) The polar points $(-1, \pi/4)$, $(-1, 9\pi/4)$, and $(1, 5\pi/4)$ all represent the Cartesian point $(-\sqrt{2}/2, -\sqrt{2}/2)$.

    (c) $(\sqrt{2}, \pi/4 + 2k\pi)$ and $(-\sqrt{2}, \pi/4 + (2k-1)\pi)$.

26. (a)

θ	0	$\frac{\pi}{6}$	$\frac{\pi}{3}$	$\frac{\pi}{2}$	$\frac{2\pi}{3}$	$\frac{5\pi}{6}$	π	$\frac{7\pi}{6}$	$\frac{4\pi}{3}$	$\frac{3\pi}{2}$	$\frac{5\pi}{3}$	$\frac{11\pi}{6}$	2π
r	2	2.5	2.866	3	2.866	2.5	2	1.5	1.134	1	1.134	1.5	2

    (c) The limaçon is symmetric with respect to the y-axis.

    (d) The r-values are symmetric about $\theta = \pi/2$.

27. (a)

θ	0	$\frac{\pi}{6}$	$\frac{\pi}{3}$	$\frac{\pi}{2}$	$\frac{2\pi}{3}$	$\frac{5\pi}{6}$	π	$\frac{7\pi}{6}$	$\frac{4\pi}{3}$	$\frac{3\pi}{2}$	$\frac{5\pi}{3}$	$\frac{11\pi}{6}$	2π
r	2	1.866	1.5	1	0.5	0.134	0	0.134	0.5	1	1.5	1.866	2

    (c) The cardioid is symmetric with respect to the x-axis.

    (d) The entries in the table are symmetric about $\theta = \pi$.

28. (a)

θ	0	$\frac{\pi}{6}$	$\frac{\pi}{3}$	$\frac{\pi}{2}$	$\frac{2\pi}{3}$	$\frac{5\pi}{6}$	π	$\frac{7\pi}{6}$	$\frac{4\pi}{3}$	$\frac{3\pi}{2}$	$\frac{5\pi}{3}$	$\frac{11\pi}{6}$	2π
r	-1	-0.732	0	1	2	2.732	3	2.732	2	1	0	-0.732	-1

    (c) $r = 0$ when $\theta = \pi/3$ and $\theta = 5\pi/3$; the graph crosses the origin at these points.

    (d) $r < 0$ when $0 \le \theta < \pi/3$ and when $5\pi/3 < \theta \le 2\pi$; these two intervals represent the lower and upper halves, respectively, of the small "loop" to the left of the origin.

49. Substituting the relations $\cos\theta = x/r$ and $\sin\theta = y/r$ into the equation $r = a\cos\theta + b\sin\theta$ produces $r = ax/r + by/r$. Multiplying both sides of this equation by r leads to $r^2 = ax + by$. Therefore,
$$x^2 + y^2 = ax + by \implies (x^2 - ax) + (y^2 - by) = 0 \implies (x - a/2)^2 + (y - b/2)^2 = (a^2 + b^2)/4.$$
From this it follows that the circle has radius $\sqrt{a^2 + b^2}/2$ and center $(a/2, b/2)$.

50. The graph of the polar equation $r = g(\theta)$ can be obtained from the graph of $r = f(\theta)$ by rotating the latter graph clockwise through an angle of α.

51. (b) Since $r = \sqrt{x^2 + y^2}$ and $\cos\theta = x/r = x/\sqrt{x^2 + y^2}$,
$r = \cos^2\theta \implies \sqrt{x^2 + y^2} = x^2/(x^2 + y^2)$ or, equivalently, $(x^2 + y^2)^{3/2} = x^2$. Squaring both sides of this equation leads to the more appealing equation $(x^2 + y^2)^3 = x^4$.

52. (b) $r = 1/(1 + \cos\theta) \implies r + r\cos\theta = 1 \implies \sqrt{x^2 + y^2} + x = 1 \implies \sqrt{x^2 + y^2} = 1 - x$. Solving the last of these equations yields $y = \sqrt{1 - 2x}$.

53. The distance between the points (x_1, y_1) and (x_2, y_2) is $\sqrt{(x_2 - x_1)^2 + (y_2 - y_1)^2}$. Using $x_1 = r_1 \cos \theta_1$, $x_2 = r_2 \cos \theta_2$, etc.,

$$\begin{aligned}
(x_2 - x_1)^2 + (y_2 - y_1)^2 &= (r_2 \cos \theta_2 - r_1 \cos \theta_1)^2 + (r_2 \sin \theta_2 - r_1 \sin \theta_1)^2 \\
&= r_2^2 \cos^2 \theta_2 + r_1^2 \cos^2 \theta_1 - 2r_1 r_2 \cos \theta_1 \cos \theta_2 \\
&\quad + r_2^2 \sin^2 \theta_2 + r_1^2 \sin^2 \theta_1 - 2r_1 r_2 \sin \theta_1 \sin \theta_2 \\
&= r_1^2 + r_2^2 - 2r_1 r_2 (\cos \theta_1 \cos \theta_2 + \sin \theta_1 \sin \theta_2) \\
&= r_1^2 + r_2^2 - 2r_1 r_2 \cos(\theta_1 - \theta_2).
\end{aligned}$$

Thus, the distance between the points (x_1, y_1) and (x_2, y_2) is $\sqrt{r_1^2 + r_2^2 - 2r_1 r_2 \cos(\theta_1 - \theta_2)}$.

54. (a) $a > 1$.

(b) When $a = 1$ the limaçon has a "dimple" rather than an inner loop.

(c) As $a \to 0$, the graph becomes a circle of radius 1 centered at the origin.

(d) The graphs are reflections of each other across the y-axis.

(e) As $a \to \infty$, the graph becomes circular.

§V.3   Calculus in Polar Coordinates

1. The results are straightforward applications of the product rule: $(f \cdot g)' = f' \cdot g + f \cdot g'$.

2. $\dfrac{dy}{dx} = \dfrac{-\sin^2\theta + (1 + \cos\theta)\cos\theta}{-\sin\theta\cos\theta - (1 + \cos\theta)\sin\theta}$

   $= \dfrac{\sin^2\theta - \cos^2\theta - \cos\theta}{(2\cos\theta + 1)\sin\theta} = \dfrac{1 - 2\cos^2\theta - \cos\theta}{(2\cos\theta + 1)\sin\theta}$

   $= \dfrac{(1 - 2\cos\theta)(1 + \cos\theta)}{(2\cos\theta + 1)\sin\theta}.$

3. (b) $\dfrac{dy}{dx} = \dfrac{\sin\theta + \theta\cos\theta}{\cos\theta - \theta\sin\theta}$. Thus, the spiral has a horizontal tangent line wherever $\theta = -\tan\theta$.

   (c) The formula for dy/dx in part (a) implies that the spiral has a vertical tangent line wherever $\theta = \cot\theta$.

   (d) The polar point $(1, 1)$ is the point $(\cos 1, \sin 1)$ in Cartesian coordinates. The slope of the tangent line at the polar point $(1, 1)$ is
   $m = (\sin 1 + \cos 1)/(\cos 1 - \sin 1) \approx -4.588$. Thus, the equation of the desired tangent line is $y = m(x - \cos 1) + \sin 1 \approx -4.588(x - 0.540) + 0.841$.

4. (b) $\dfrac{dy}{dx} = \dfrac{2\cos\theta\sin\theta + (1 + 2\sin\theta)\cos\theta}{2\cos^2\theta - (1 + 2\sin\theta)\sin\theta} = \dfrac{4\cos\theta\sin\theta + \cos\theta}{2\cos^2\theta - 2\sin^2\theta - \sin\theta}$. Thus, at the
   polar point $(1, 0)$ the limaçon has slope $1/2$. Thus, the tangent line is described by the equation $y = (x - 1)/2$.

   (c) The polar point $(0, 7\pi/6)$ corresponds to the Cartesian point $(0, 0)$. The slope of the tangent line at this point is $\sqrt{3}/3$, so the equation of the tangent line is $y = \sqrt{3}x/3$.

   (d) The polar point $(0, 11\pi/6)$ corresponds to the Cartesian point $(0, 0)$. The slope of the tangent line at this point is $-\sqrt{3}/3$, so the equation of the tangent line is $y = -\sqrt{3}x/3$.

   (e) Thus, the limaçon has horizontal tangent lines when $\cos\theta = 0$ or $\sin\theta = -1/4$—at the points $(3, \pi/2)$, $(-1, 3\pi/2)$, $(1/2, 3.3943)$, and $(1/2, 6.0305)$.

5. area $= \dfrac{1}{2}\displaystyle\int_0^\pi 1\, d\theta = \pi/2.$

6. area $= \dfrac{1}{2}\displaystyle\int_0^{\pi/2} 9\, d\theta = \dfrac{9\pi}{4}.$

7. area $= \dfrac{1}{2}\displaystyle\int_0^\pi a^2\, d\theta = \dfrac{\pi a^2}{2}.$

8. area $= \dfrac{1}{2}\displaystyle\int_0^\beta 1\, d\theta = \dfrac{\beta}{2}.$

9. area $= \int_0^{\pi/2} f(\theta)^2 \, d\theta + \int_{11\pi/6}^{2\pi} f(\theta)^2 \, d\theta = \frac{1}{2} \int_{-\pi/6}^{7\pi/6} f(\theta)^2 \, d\theta = 2\pi + \frac{3\sqrt{3}}{2}.$

10. area $= \frac{1}{2} \int_{2\pi/3}^{4\pi/3} f(\theta)^2 \, d\theta = \frac{1}{2} \int_{2\pi/3}^{4\pi/3} (1 + 2\cos\theta)^2 \, d\theta$

    $= \left(\frac{3\theta}{2} + 2\sin\theta + \cos\theta\sin\theta \right)\bigg]_{2\pi/3}^{4\pi/3} = \pi - 3\sqrt{3}/2.$

11. area $= \frac{1}{2} \int_{-\pi/2}^{\pi/2} f(\theta)^2 \, d\theta = \frac{1}{2} \int_{-\pi/2}^{\pi/2} (1 - \cos\theta)^2 \, d\theta = \int_0^{\pi/2} (1 - \cos\theta)^2 \, d\theta$

    $= \left(\frac{3\theta}{2} - 2\sin\theta + \frac{1}{2}\cos\theta\sin\theta \right)\bigg]_0^{\pi/2} = 3\pi/4 - 2.$

12. area $= \frac{1}{2} \int_{\pi/2}^{\pi} f(\theta)^2 \, d\theta = \frac{1}{2} \int_{\pi/2}^{\pi} (1 - \cos\theta)^2 \, d\theta$

    $= \left(\frac{3\theta}{4} - \sin\theta + \frac{\cos\theta\sin\theta}{4} \right)\bigg]_{\pi/2}^{\pi} = 1 + 3\pi/8.$

13. area $= \frac{1}{2} \int_0^{\pi/2} \left(f(\theta)^2 - g(\theta)^2 \right) d\theta + \frac{1}{2} \int_{\pi}^{2\pi} f(\theta)^2 \, d\theta$

    $= \frac{1}{2} \int_0^{\pi/2} \left((1 + \cos\theta)^2 - \sin^2\theta \right) d\theta + \frac{1}{2} \int_{\pi}^{2\pi} (1 + \cos\theta)^2 \, d\theta$

    $= \left(\frac{\theta}{2} + \sin\theta + \frac{\cos\theta\sin\theta}{2} \right)\bigg]_0^{\pi/2} + \left(\frac{3\theta}{4} + \sin\theta + \frac{\cos\theta\sin\theta}{4} \right)\bigg]_{\pi}^{2\pi}$

    $= (1 + \pi/4) + 3\pi/4 = 1 + \pi.$

14. area $= \frac{1}{2} \int_0^{\pi/4} \sec^2\theta \, d\theta = \frac{1}{2}\tan\theta\bigg]_0^{\pi/4} = \frac{1}{2}.$

15. area $= m/2.$

    area $= \frac{1}{2} \int_0^{\arctan m} \sec^2\theta \, d\theta = \frac{1}{2}\tan\theta\bigg]_0^{\arctan m} = \frac{m}{2}.$

16. (a) One leaf lies between $\theta = -\pi/2n$ and $\theta = \pi/2n$. The area of this leaf is
    $\frac{1}{2} \int_{-\pi/2n}^{\pi/2n} \cos^2(n\theta) \, d\theta = \pi/4n.$

    (b) The area of all $2n$ leaves is $\pi/2$ (i.e., one-half of the area of the circle $r = 1$).

17. (a) One leaf lies between $\theta = -\pi/2n$ and $\theta = \pi/2n$. The area of this leaf is
    $\frac{1}{2} \int_{-\pi/2n}^{\pi/2n} \cos^2(n\theta) \, d\theta = \pi/4n.$

    (b) The area of all n leaves is $\pi/4$ (i.e., one-fourth of the area of the circle $r = 1$).

18. area $= \dfrac{1}{2} \displaystyle\int_0^{2\pi} \theta^2 \, d\theta = 4\pi^3/3$.

19. area $= \left(e^{4\pi} - 1\right)/4$.

20. area $= \displaystyle\int_{2\pi}^{4\pi} \dfrac{\ln(\theta)^2}{2} \, d\theta \approx 15.664$. (The integral can be done in closed form, but it is very messy.)

21. area $= \dfrac{1}{2} \displaystyle\int_{-\pi/3}^{\pi/3} d\theta - \dfrac{1}{2} \int_{-\pi/3}^{\pi/3} \left(\tfrac{1}{2} \sec\theta\right)^2 d\theta = \dfrac{\pi}{3} - \dfrac{\sqrt{3}}{4}$.

22. area $= \dfrac{1}{2} \displaystyle\int_{-\arccos a}^{\arccos a} d\theta - \dfrac{1}{2} \int_{-\arccos a}^{\arccos a} (a \sec\theta)^2 \, d\theta$

    $= \arccos a - a^2 \tan(\arccos a) = \arccos a - a\sqrt{1 - a^2}$.

23. $x = 2\cos t,\ y = 2\sin t,\ 0 \le t \le 2\pi$. The graph is a circle of radius 2 with center at the origin.

24. $x = 2\cos t,\ y = 2\sin t,\ 0 \le t \le \pi$. The graph is the upper half of a circle of radius 2 with center at the origin.

25. $x = 1,\ y = \tan t,\ -\pi/4 \le t \le \pi/4$. The graph is the vertical line segment from $(1, -1)$ to $(1, 1)$.

26. $x = \cot t,\ y = 1,\ \pi/4 \le t \le 3\pi/4$. The graph is the horizontal line segment from $(1, 1)$ to $(-1, 1)$.

27. $x = t\cos t,\ y = t\sin t,\ 0 \le t \le 2\pi$. The graph is one loop of a spiral.

28. $x = \cos^2 t,\ y = \cos t \sin t,\ 0 \le t \le \pi$. The graph is a circle of radius $1/2$ with center at $(1/2, 0)$.

29. $x = \cos(2t)\cos t,\ y = \cos(2t)\sin t,\ 0 \le t \le 2\pi$. The graph is the 4-leaf rose shown in Example 3.

30. $x = \cos t + \cos^2 t,\ y = \sin t + \cos t \sin t,\ 0 \le t \le 2\pi$. The graph is the cardioid shown in Example 1.

31. $\left(\dfrac{dy}{d\theta}\right)^2 + \left(\dfrac{dx}{d\theta}\right)^2 = \left(f'(\theta)\sin\theta + f(\theta)\cos\theta\right)^2 + \left(f'(\theta)\cos\theta - f(\theta)\sin\theta\right)^2$

$\qquad = \left(f'(\theta)^2 \sin^2\theta + 2f(\theta)f'(\theta)\sin\theta\cos\theta + f(\theta)^2 \cos^2\theta\right)$

$\qquad\qquad + \left(f'(\theta)^2 \cos^2\theta - 2f(\theta)f'(\theta)\sin\theta\cos\theta + f(\theta)^2 \sin^2\theta\right)$

$\qquad = f'(\theta)^2 + f(\theta)^2 > 0$,

which implies that $dx/d\theta$ and $dy/d\theta$ are not both simultaneously zero.

32.   (a) The graph of $1 + a \cos \theta$ has a "dimple" if $|a| > 1/2$, and has no dimple if $|a| \leq 1/2$.

      (b) $\dfrac{dy}{dx} = \dfrac{-a \sin^2 \theta + \cos \theta + a \cos^2 \theta}{-2a \sin \theta \cos \theta - \sin \theta}$. The limaçon has a vertical tangent whenever the denominator in this expression is zero (i.e., when $\theta = 0$ or $\cos \theta = -1/(2a)$). Thus, there will be three vertical tangent lines if and only if $|a| \leq 1/2$.

M   *Multivariable Calculus: A First Look*

§M.1   Three-Dimensional Space

1. (a) The surface is unrestricted in the x-direction.

    (b) The surface is unrestricted in the y-direction.

2. The graph in the xy-plane is a circle; in xyz-space it's a pipe centered on the z-axis.

3. Two planes, crossing diagonally along the x-axis.

4. The x-axis.

5. A parabolic "tent," unrestricted in the x-direction.

6. The empty set, because the equation $y^2 + z^2 = -1$ has no solutions.

7. $x^2 + y^2 + z^2 = 4$.

8. $(x - 1)^2 + (y - 1)^2 + (z - 1)^2 = 1$.

9. $x^2 + z^2 = 1$.

10. $x^2 + y^2 = 4$.

11. $z = \sin x$ and $z = \cos x$ are two possibilities.

12. Yes, the graph is "cylindrical" in both the x-direction and the y-directions. The trace in the xz-plane (where $y = 0$) is the line $z = 3$. The trace in the yz-plane (where $x = 0$) is the line $z = 3$. The trace in the xy-plane (where $z = 0$) is empty.

13. Completing the square in y and z gives
    $x^2 + y^2 - 6y + z^2 - 4z = 0 \iff x^2 + (y - 3)^2 + (z - 2)^2 = 13$. Thus we have a sphere of radius $\sqrt{13}$, centered at $(0, 3, 2)$.

14. (a) Setting $x = 0$ in $x^2 + y^2 + z^2 = 1$ gives $y^2 + z^2 = 1$—the unit circle in the yz-plane.

    (b) Setting $y = 0$ in $x^2 + y^2 + z^2 = 1$ gives $x^2 + z^2 = 1$—the unit circle in the xz-plane.

    (c) Draw half circles in each of the three coordinate planes; join them to get part of the sphere.

    (d) Setting $z = 1/2$ in $x^2 + y^2 + z^2 = 1$ gives $x^2 + y^2 = 3/4$—the circle of radius $\sqrt{3}/2$ centered at $(0, 0)$ in the plane $z = 1/2$.

    (e) Setting $z = a$ in $x^2 + y^2 + z^2 = 1$ gives $x^2 + y^2 = 1 - a^2$—the circle of radius $\sqrt{1 - a^2}$ centered at $(0, 0)$ in the plane $z = a$. Thus, the trace of S in the plane $z = 0.9$ is the circle of radius $\sqrt{1 - 0.9^2} = \sqrt{0.19}$ centered at the origin. The trace of S in the plane $z = 1$ is the origin. If $a = 2$, there is no intersection (i.e., the trace does not exist).

15. (b) According to the Pythagorean rule, $|OQ|^2 = |OP|^2 + |PQ|^2 = 1 + 4 = 5$ so the length of OQ is $\sqrt{5}$.

(c) According to the Pythagorean rule, $|OR|^2 = |OQ|^2 + |QR|^2 = 5 + 9 = 14$ so the length of QR is $\sqrt{14}$.

(d) Using the distance formula, $|OQ| = \sqrt{(1-0)^2 + (2-0)^2} = \sqrt{5}$ and $|OR| = \sqrt{(1-0^2 + (2-0)^2 + (3-0)^2} = \sqrt{14}$.

16. $d(P, Q) = \sqrt{(a-x)^2 + (b-y)^2 + (c-z)^2}$. Since $P \neq Q$, $(a-x)^2 + (b-y)^2 + (c-z)^2 > 0$. It follows that $d(P, Q) > 0$.

17. $d(P, Q) = \sqrt{(a-x)^2 + (b-y)^2 + (c-z)^2}$
$$= \sqrt{(x-a)^2 + (y-b)^2 + (z-c)^2} = d(Q, P).$$

18. Let $M = \big((x+a)/2, (y+b)/2, (z+c)/2\big)$. Then,

$$d(P, M) = \sqrt{\big((x+a)/2 - x\big)^2 + \big((y+b)/2 - y\big)^2 + \big((z+c)/2 - z\big)^2}$$
$$= \sqrt{(a-x)^2/4 + (b-y)^2/4 + (c-z)^2/4}$$
$$= \left(\sqrt{(a-x)^2 + (b-y)^2 + (c-z)^2}\right)/2$$
$$= d(P, Q)/2.$$

Similarly,

$$d(M, Q) = \sqrt{\big(a - (x+a)/2\big)^2 + \big(b - (y+b)/2\big)^2 + \big(c - (z+c)/2\big)^2}$$
$$= \sqrt{(a-x)^2/4 + (b-y)^2/4 + (c-z)^2/4}$$
$$= \left(\sqrt{(a-x)^2 + (b-y)^2 + (c-z)^2}\right)/2$$
$$= d(P, Q)/2.$$

19. $(1, 1, 1), (1, 1, -1), (1, -1, 1), (1, -1, -1), (-1, 1, 1), (-1, 1, -1), (-1, -1, 1),$ $(-1, -1, -1)$.

20. If $A = B = 0$, then $Ax + By = C$ becomes $0 = C$, which is either trivial or impossible.

21. The line $Ax + By = C$ has slope $-A/B$, so lines with $B = 0$ have undefined slope.

22. Setting $x = 0$ in $Ax + By = C$ gives $By = C$, or $y = C/B$. Thus, lines with $B = 0$ have no y-intercept.

23. Setting $y = 0$ in $Ax + By = C$ gives $Ax = C$, or $x = C/A$. Thus, lines with $A = 0$ have no x-intercept.

24. If $A = B = C = 0$, then the equation becomes $0 = D$, which is either trivial or impossible.

25. Setting $y = 0$ and $z = 0$ in $Ax + By + Cz = D$ gives $Ax = D$, so the x-intercept is $x = D/A$. Thus any plane with $A = 0$ and $D \neq 0$ (such as $y + z = 1$) has no x-intercept.

26. Setting $x = 0$ and $y = 0$ in $Ax + By + Cz = D$ gives $Cz = D$, so the z-intercept is $z = D/C$. Thus any plane with $C = 0$ and $D \neq 0$ (such as $x + y = 1$) has no z-intercept.

27. (a) Setting $y = 0$ in $x + 2y + 3z = 3$ gives the trace $x + 3z = 3$; this line intercepts the x- and z-axes at $x = 3$ and $z = 1$, respectively.

    (b) Setting $x = 0$ in $x + 2y + 3z = 3$ gives the trace $2y + 3z = 3$; this line intercepts the y- and z-axes at $y = 3/2$ and $z = 1$, respectively.

    (c) Setting $x = 1$ in $x + 2y + 3z = 3$ gives the trace $1 + 2y + 3z = 3$, or $2y + 3z = 2$.

28. (a) Setting $y = 0$ and $z = 0$ gives $x = 1$—that's the x-intercept. Similarly, $y = 2$ and $z = 4$ are the y- and z-intercepts. Connecting these points in the first quadrant gives a picture of the plane.

    (b) The traces are $2y + z = 4$ in the yz-plane; $4x + z = 4$ in the xz-plane; $4x + 2y = 4$ in the xy-plane. These are the lines joining the points found in part (a).

29. The line $1x + 0y = 3$ intercepts the x-axis but not the y-axis.

30. The plane $x = 1$ in xyz-space intercepts only the x-axis, at the point $x = 1$.

31. The plane $x + 2y = 1$ in xyz-space intercepts the x-axis at $x = 1$ and the y-axis at $y = 1/2$.

32. Any plane of the form $bz + cy = d$, where all coefficients are non-zero, hits the y-axis and the z-axis but not the x-axis.

33. The midpoint has coordinates $((x_1 + x_2)/2, (y_1 + y_2)/2, (z_1 + z_2)/2)$; one shows that these coordinates satisfy the equation of the given plane.

§M.2   Functions of Several Variables

1. $g(x, y) = x^2 + y^2$ has domain $\mathbb{R}^2$; the range is $[0, \infty)$.

2. $h(x, y) = x^2 + y^2 + 3$ has domain $\mathbb{R}^2$; the range is $[3, \infty)$.

3. $j(x, y) = 1/(x^2 + y^2)$ has domain all of $\mathbb{R}^2$ except for the origin; the range is $(0, \infty)$.

4. $k(x, y) = x^2 - y^2$ has domain $\mathbb{R}^2$; the range is $(-\infty, \infty)$.

5. The domain of $m(x, y) = \sqrt{1 - x^2 - y^2}$ is all points inside and on the unit circle in $\mathbb{R}^2$; the range is $[0, 1]$.

6. Always, $f(x, y) = y - x^2$ and $g(x, y) = x - y^2$.

    (a) Level curves are a family of parabolas.

    (b) Level curves are a family of parabolas.

    (c) Both are families of parabolas. The level curves of f open upward and are symmetric about the y-axis. The level curves of g open to the right and are symmetric about the x-axis.

    (d) The two surfaces are the same except for their orientation: f is a curved surface that arches over the y-axis while g arches over the x-axis.

7. (a) In the rectangle All level curves of f are circles centered at the origin, except the curve $z = 0$, which is just the origin.

    (b) All level curves of g are circles centered at the origin, except for except the curve $z = 1$, which is just the origin.

    (c) The curves themselves are identical, but their labels are different.

    (d) Both graphs are paraboloids, opening upward. The g-graph is one unit higher.

8. (a) All the level curves are straight lines with slope 2/3.

    (b) These level curves are also straight lines—the same lines as in the previous part.

    (c) The level lines are the same for both functions, but their labels are different. Specifically, the level line $z = c$ for f is the level line $z = -c$ for g.

    (d) All level curves of linear functions are straight lines.

    (e) Both graphs are planes. At each (x, y), the two planes have opposite z-coordinates.

9. (a) Each level curve is a vertical line (since the value of f depends only on x).

    (b) Each level curve is a horizontal line (since the value of g depends only on y).

    (c) In both parts, the level curves are lines. However, the level curves of f are vertical lines while the level curves of g are horizontal lines.

    (d) The graph of g can be obtained from the graph of f by rotating it by $\pi/2$ radians.

   (e) The contour map of a cylinder is a horizontal line or a vertical line (depending on which variable is unrestricted).

10. In weather language, $T(0, 0) = 15$ means that the temperature was 15 degrees Celsius at noon CST in Los Angeles on January 1, 2001.

11. Level curves of T, in weather language, are curves along which the temperature remains constant. As a rule, such "isotherms" run east-and-west.

12. $T(1400, 1100) = -15$

13. Near the coldest (or warmest) spot in the country, the level curves should be closed curves, shrinking down to the cold spot.

14.   (a) The domain and range of f are $\mathbb{R}^2$ and $[0, \infty)$, respectively.

   (b) The graph of f is a cone in xyz-space.

   (c) The level curve of f through $(3, 4)$ is a circle of radius 5, centered at the origin.

   (d) All level curves of f are circles centered at the origin.

15.   (a) $f(x, y) = |x - 1|$

   (b) The graph of f is a vee-shaped trough, parallel to the y-axis.

   (c) The level curve of f through $(3, 4)$ is the straight line $x = 3$.

   (d) All level curves of f are straight lines parallel to the y-axis.

16.   (a) The level curve $z = c$ is a straight lines of the form $3x - 2y = c$. The lines are equally spaced, all with slope 2/3.

   (b) The level line $z = 0$ has equation $3x - 2y = 0$.

   (c) The general formula is $L(x, y) = 3x - 2y$.

   (d) The graph of L is a plane; this shape is indeed consistent with the level curves plotted earlier.

§M.3 Partial Derivatives

1. $f_x(x, y) = 2x$; $f_y(x, y) = -2y$.

2. $f_x(x, y) = 2xy^2$; $f_y(x, y) = 2x^2y$.

3. $f_x(x, y) = \dfrac{2x}{y^2}$; $f_y(x, y) = \dfrac{-2x^2}{y^3}$.

4. $f_x(x, y) = -y\sin(xy)$; $f_y(x, y) = -x\sin(xy)$.

5. $f_x(x, y) = -\sin(x)\cos(y)$; $f_y(x, y) = -\cos(x)\sin(y)$.

6. $f_x = \dfrac{-\sin(x)}{\cos(y)}$; $f_y = \dfrac{\cos(x)\sin(y)}{\cos^2(y)}$.

7. $f_x(x, y, z) = y^2z^3$; $f_y(x, y, z) = 2xyz^3$; $f_z(x, y, z) = 3xy^2z^2$.

8. $f_x(x, y, z) = -yz\sin(xyz)$; $f_y(x, y, z) = -xz\sin(xyz)$; $f_z(x, y, z) = -xy\sin(xyz)$.

9. (a) Because $f(3) = 9$ and $f'(3) = 6$, it follows that the linear approximation function has equation $L(x) = 9 + 6(x - 3)$.

   (b) The graphs of f and L are close together near $x = 3$; in particular, the graph of f looks similar to a straight line.

   (c) The inequality $|f(x) - L(x)| < 0.01$ holds for x in the interval $2.9 < x < 3.1$. This can be found either graphically, by zooming, or algebraically. The latter approach can be done as follows. Notice first that

   $$|f(x) - L(x)| = |x^2 - (9 + 6(x - 3))| = |x^2 - 6x + 9| = |x - 3|^2.$$

   Thus

   $$|f(x) - L(x)| < 0.01 \iff |x - 3|^2 < 0.01 \iff |x - 3| < 0.1.$$

   This is equivalent to saying that $2.9 < x < 3.1$.

10. (a) Since $f(9) = 3$ and $f'(3) = 1/6$, it follows that the linear approximation function L has formula $L(x) = 3 + (x - 9)/6$.

    (b) Plotting f and L in the vicinity of $x = 9$ (e.g., on the interval $6 < x < 12$) shows that the two graphs are very close together.

    (c) Looking carefully at the graphs of f and L near $x = 9$ shows that the functions differ by less than 0.01 on the interval $8 < x < 10$. (The exercise can also be solved algebraically, but it's rather messy.)

11. (a) The fact that $f'(3) = 6$ means that the graph of f looks like a line of slope 6 near $x = 3$. (The graph bears this out.)

(b) Both of the secant lines, from $x = 3$ to $x = 3.5$ and from $x = 3$ to $x = 3.1$, have slopes near 6.

(c) The average rates of change $\Delta y / \Delta x$ of f over the intervals $[3, 3.5]$ and $[3, 3.1]$ are 6.5 and 6.1, respectively.

(d) For $f(x) = x^2$, $\lim\limits_{h \to 0} \dfrac{f(3+h) - f(3)}{h} = 6$. The answer is the derivative of f at $x = 3$.

12. (a) The fact that $f'(3) = 5$ means that the graph of f looks like a line of slope 5 near $x = 3$. (The graph bears this out.)

   (b) Both of the secant lines, from $x = 3$ to $x = 3.5$ and from $x = 3$ to $x = 3.1$, have slopes near 5.

   (c) The average rates of change $\Delta y / \Delta x$ of f over the intervals $[3, 3.5]$ and $[3, 3.1]$ are 5.5 and 5.1, respectively.

   (d) For $f(x) = x^2 - x$, $\lim\limits_{h \to 0} \dfrac{f(3+h) - f(3)}{h} = 5$. The answer is the derivative of f at $x = 3$.

13. $g_x(1, 1) \approx (g(1.01, 1) - g(1, 1))/0.01 = -2.01$. Similarly, $g_y(1, 1) \approx (g(1, 1.01) - g(1, 1))/0.01 = 3.01$.

14. The table reflects the fact that $g_x(0, 0) = 0$ in that the entries along the *row* through the $(0, 0)$-position are close to constant. The table reflects the fact that $g_y(0, 0) = 1$ in that the entries along the *column* through the $(0, 0)$-position vary at the same rate as y.

15. (a) Since $L(x, y) = y$, $L_x(x, y) = 0$ and $L_y(x, y) = 1$. It follows that $L(0, 0) = 0$, $L_x(0, 0) = 0$, and $L_y(0, 0) = 1$.

   (b)

y \ x	−0.02	−0.01	0.00	0.01	0.02
0.02	0.02	0.02	0.02	0.02	0.02
0.01	0.01	0.01	0.01	0.01	0.01
0.00	0.00	0.00	0.00	0.00	0.00
−0.01	−0.01	−0.01	−0.01	−0.01	−0.01
−0.02	−0.02	−0.02	−0.02	−0.02	−0.02

16. To find the linear approximation function $M(x, y)$ with (i) $M(1, 1) = g(1, 1)$; (ii) $M_x(1, 1) = g_x(1, 1)$; (iii) $M_y(1, 1) = g_y(1, 1)$, we need to know or estimate the right-hand quantities. We know $g(1, 1) = 1$ from the table, and we estimated $g_x(1, 1) \approx -2$ and $g_y(1, 1) \approx 3$. This gives $M(x, y) = 1 - 2(x - 1) + 3(y - 1)$ as a reasonable estimate. (In fact, $g(x, y) = y^2 + y - x^2$ so $g_y = 2y + 1$ and $g_x = -2x$. But these formulas aren't necessary for doing the exercise.)

17. (a) The graph is a cylinder in the x-direction.

   (b) $f_x(x, y) = 0$; $f_y(x, y) = \cos y$. The fact that $f_x = 0$ reflects the fact that f is constant in x.

(c) The linear approximation function L for f at the point $(0, 0)$ is $L(x, y) = y + 2$. Like f itself, L is independent of x.

18. (a) Mimicking the reasoning in Example 3, (page M-20), gives $f_x(1.5, 1.5) \approx -1.5$ and $f_y(1.5, 1.5) \approx -4.5$.

    (b) The formulas $f_x = 2x - 3y$ and $f_y = -3x$ show that $f_x(1.5, 1.5) = -1.5$ and $f_y(1.5, 1.5) = -4.5$.

    (c) It follows from part (b), and the fact that $f(1.5, 1.5) = 1.5$, that
    $$L(x, y) = -1.5(x - 1.5) - 4.5(y - 1.5) + 1.5 = -1.5x - 4.5y + 10.5.$$

19. (a) All the level curves are vertical lines.

    (b) >From the diagram, $f_x(0, 0) \approx 1$ and $f_y(0, 0) = 0$.

    (c) >From the diagram, $f_x(\pi/2, 0) \approx 0$ and $f_y(0, 0) = 0$.

    (d) All contour lines are vertical; this means that $f(x, y)$ is constant along vertical lines, which implies that $f_y(x, y) = 0$.

    (e) The fact that $f_x(x, y)$ is independent of y means that the rate of increase of f in the x-direction is the same for all y. The contour map of f reflects this fact in that contour lines are all vertical.

20. (a) All the level curves are horizontal lines.

    (b) From the diagram, $f_x(0, 0) = 0$ and $f_y(0, 0) \approx 0$.

    (c) From the diagram, $f_x(\pi/2, 0) = 0$ and $f_y(\pi/2, 0) \approx -1$.

    (d) All contour lines are horizontal; this means that $f(x, y)$ is constant along horizontal lines, which implies that $f_x(x, y) = 0$.

    (e) The fact that $f_y(x, y)$ is independent of x means that the rate of increase of f in the y-direction is the same for all x. The contour map of f reflects this fact in that contour lines are all horizontal.

21. (a) All the level curves are straight lines; all have the same slope, and they're equally spaced.

    (b) The contour map shows that $f_x(0, 0) = 2$ and $f_y(0, 0) = -3$; one sees this by measuring how fast f increases in the x- and y-directions.

    (c) The contour map of f reflects the fact that both f_x and f_y are constant functions in the sense that all contour lines are straight and equally spaced.

    (d) The contour map shows that $f_x(x, y)$ is positive and $f_y(x, y)$ is negative in that moving to the right (in the x-direction) corresponds to increasing z, while moving upward (in the y-direction) corresponds to decreasing z.

22. (a) All the level curves are straight lines; all have the same slope, and they're equally spaced.

(b) The contour map shows that $f_x(0, 0) = -1$ and $f_y(0, 0) = 2$; one sees this by measuring how fast f increases in the x- and y-directions.

(c) The contour map shows that $f_x(x, y)$ is negative and $f_y(x, y)$ is positive in that moving to the right (in the x-direction) corresponds to decreasing z, while moving upward (in the y-direction) corresponds to increasing z.

23. (a) The partial derivatives of $f(x, y) = xy$ are $f_x(x, y) = y$ and $f_y(x, y) = x$. Thus, at the base point $(x_0, y_0) = (2, 1)$, $f(2, 1) = 2$, $f_x(2, 1) = 1$, and $f_y(2, 1) = 2$. Therefore $L(x, y) = 1(x - 2) + 2(y - 1) + 2 = x + 2y - 2$.

(b) Level curves of $f(x, y) = xy$ are hyperbolas, of the form $xy = k$. Level curves of $L(x, y) = x + 2y - 2$ are straight lines, of the form $x + 2y - 2 = k$.

(c) The level curves of f and L look similar near $(2, 1)$. (E.g., all the level curves of L are lines, with slope $-1/2$. Similarly, the hyperbola $xy = 2$ (a level curve of f) has slope $-1/2$ at the point $(2, 1)$.

(d) In a small window around $(2, 1)$, the level curves of f and L appear almost identical—this is because L is the linear approximation to f at $(2, 1)$.

24. (a) The partial derivatives of $f(x, y) = x^2 - y^2$ are $f_x(x, y) = 2x$ and $f_y(x, y) = -2y$. Thus, at the base point $(x_0, y_0) = (2, 1)$, $f(2, 1) = 3$, $f_x(2, 1) = 4$, and $f_y(2, 1) = -2$. Therefore $L(x, y) = 4(x - 2) - 2(y - 1) + 3 = 4x - 2y - 3$.

(b) Level curves of $f(x, y) = x^2 - y^2$ are hyperbolas, of the form $x^2 - y^2 = k$. Level curves of $L(x, y) = 4x - 2y - 3$ are straight lines, of the form $4x - 2y - 3 = k$.

(c) The level curves of f and L look similar near $(2, 1)$. (E.g., all the level curves of L are lines, with slope 2. Similarly, the hyperbola $x^2 - y^2 = 3$ (a level curve of f) has slope 2 at the point $(2, 1)$.

(d) In a small window around $(2, 1)$, the level curves of f and L appear almost identical—this is because L is the linear approximation to f at $(2, 1)$.

25. (a) From the contour map of f we estimate $f_x(1, 2) = 2$ and $f_y(1, 2) = 4$.

(b) Symbolic differentiation of $f(x, y) = x^2 + y^2$ gives $f_x = 2x$ and $f_y = 2y$, so $f_x(1, 2) = 2$ and $f_y(1, 2) = 4$.

(c) It follows from the above and from the fact that $f(1, 2) = 5$ that $L(x, y) = 2(x - 1) + 4(y - 2) + 5 = 2x + 4y - 5$.

(d) Level curves $L(x, y) = k$ are all straight lines, with slope $-1/2$. Level curves $f(x, y) = k$ are all circles, centered at the origin. At the point $(1, 2)$, the level curves for f and L are tangent to each other.

26. (a) $f_x(x, y) = \cos x + y$; $f_y(x, y) = 2 + x$, so $f_x(0, 0) = 1$ and $f_y(0, 0) = 2$.

(b) The linear approximation to f at $(0, 0)$ is $L(x, y) = x + 2y$.

(c) All parts are routine, given the formulas. The two rows all have similar entries, especially at the left, because L approximates f closely near $(0, 0)$.

(d) Corresponding level curves of the two functions should appear similar, especially near $(0, 0)$.

27. If $f(x, y) = x^2 + y^2$ and $(x_0, y_0) = (2, 1)$, then $L(x, y) = -5 + 4x + 2y$.

28. If $f(x, y) = x^2 + y^2$ and $(x_0, y_0) = (0, 0)$, then $L(x, y) = 0$.

29. If $f(x, y) = \sin(x) + \sin(y)$ and $(x_0, y_0) = (0, 0)$, then $L(x, y) = x + y$.

30. If $f(x, y) = \sin(x) \sin(y)$ and $(x_0, y_0) = (0, 0)$, then $L(x, y) = 0$.

31. Suppose that f is independent of x. Then, $f_x(x, y) = 0$ at all points (x, y). Therefore,
$$L(x, y) = f(x_0, y_0) + f_x(x_0, y_0)(x - x_0) + f_y(x_0, y_0)(y - y_0) = f(x_0, y_0) + f_y(x_0, y_0)(y - y_0)$$
which is also a function that is independent of x.

32.   (a) The information given about partial derivatives implies that
$$L(x, y) = 6(x - 3) + 8(y - 4) + 25 = 6x + 8y - 25.$$

    (b) Use the formula above: $L(2.9, 3.1) = 6(-0.1) + 8(-0.1) + 25 = 23.6$;
$L(3.1, 4.1) = 6(0.1) + 8(0.1) + 25 = 26.4$; $L(4, 5) = 39$.

    (c) The function f can not be linear. If f were linear it would agree everywhere with L; but we saw above that $f(4, 5) \neq L(4, 5)$.

33.   (a) The information given on partial derivatives implies that
$$L(x, y) = 0.6(x - 3) + 0.8(y - 4) + 5 = 0.6x + 0.8y.$$

    (b) Use the formula above: $L(2.9, 4.1) = 0.6(-0.1) + 0.8(0.1) + 5 = 5.02$;
$L(4, 5) = 0.6(1) + 0.8(1) + 5 = 6.4$.

    (c) The function g can not be linear. If g were linear it would agree everywhere with L—but we saw above that $g(4, 5) = \sqrt{41} \neq 6.4 = L(4, 5)$.

34.   (a) The graph is "creased" along the x-axis. This suggests no y-derivative there.

    (b) The limit $\lim\limits_{h \to 0} \dfrac{f(0, h) - f(0, 0)}{h} = \lim\limits_{h \to 0} \dfrac{|h|}{h}$ doesn't exist because $\lim\limits_{h \to 0^+} \dfrac{|h|}{h} = 1$ but $\lim\limits_{h \to 0^-} \dfrac{|h|}{h} = -1$.

    (c) Since $f(x, 0) = 0$, it follows that $f_x(0, 0) = 0$. On the graph, the surface is horizontal in the x-direction at the origin.

    (d) Since $f(x, \pi/2) = 0$, it follows that $f_y(0, \pi/2) = 0$.

    (e) Zooming in on the graph near $(0, \pi/2)$ shows that at this point the surface is "flat" in the y-direction.

    (f) One possibility is $g(x, y) = |x| \cos y$.

35.   (a) Graphs show a crease along the y-axis; this suggests that f_x may not exist there. f_y exists everywhere.

    (b) Since $f(x, 0) = 0$, it follows that $f_x(x, 0) = 0$.

    (c) Since $f(x, 1) = |x|$, it follows that $f_x(x, 1)$ doesn't exist.

§M.4 Optimization and Partial Derivatives: A First Look

1. (a) The ant's altitude remains constant.

    (b) The ant rises all the way; it's highest at $(0.5, 1)$. At that point, its altitude is 0.5.

2. (a) The eggs go above the black circles.

    (b) For example, $(0, \pi/2)$ is a local maximum point, $(0, -\pi/2)$ is a local minimum point, and $(\pi/2, 0)$ is a saddle point.

    (c) Because $f_x = -\sin(x)\sin(y)$ and $f_y = \cos(x)\cos(y)$, there are four stationary points in $[-3, 3] \times [-3, 3]$: they're at $(0, \pi/2)$, $(0, -\pi/2)$, $(\pi/2, 0)$, and $(-\pi/2, 0)$.

    (d) The maximum value is 1; the minimum is -1.

3. (a) The function f has stationary points $(1, \pi/2)$ and $(1, -\pi/2)$. The first is a local minimum, the second a local maximum, apparently.

    (b) The partial derivatives are $f_x(x, y) = (2x - 2)\sin(y)$; $f_y(x, y) = (x^2 - 2x)\cos(y)$. It follows that the stationary points are at $(0, 0)$, $(2, 0)$, $(1, \pi/2)$ and $(1, -\pi/2)$.

    (c) The functions $f(-3, y) = -15\sin(y)$ has minimum value -15 at $y = -\pi/2$ and maximum value 15 at $y = \pi/2$.

4. There's a local maximum at $(0, 0)$.

5. There's a saddle at $(0, 0)$.

6. There's a local minimum at $(0, 0)$.

7. There's a saddle at $(1, 2)$.

8. The function $L(x, y) = 1 + 2x + 3y$ has no stationary points, because $L_x(x, y) = 2$ and $L_y(x, y) = 3$ for all (x, y).

9. (a) A plane can have a stationary points only if the plane is horizontal. In this case, every point is a stationary point.

    (b) $L_x = a$ and $L_y = b$, so there are stationary points only if $a = b = 0$. (c can have any value.) In this case, every point is stationary.

10. (a) All points on the y-axis (where $x = 0$) are stationary; they're local minimum points.

    (b) Since $f(x, y) = x^2$, $f_x = 2x$ and $f_y = 0$. Therefore, every point with $x = 0$ is stationary. (This agrees with part (a).)

11. $g(x, y) = y^2$ works; so do many others.

12. $h(x, y) = -(x - 1)^2$ works; so do many others.

13. $k(x, y) = (x - 3)^2 + (y - 4)^2$ works; so do many others.

§M.5 Multiple Integrals and Approximating Sums

1. Doing this by hand gives $\frac{1}{4} f\left(\frac{1}{8}\right) + \frac{1}{4} f\left(\frac{3}{8}\right) + \frac{1}{4} f\left(\frac{5}{8}\right) + \frac{1}{4} f\left(\frac{7}{8}\right)$. Using $f(x) = x^2$, and working this out as a fraction, gives 21/64.

   Maple's value for M_{100}, the midpoint sum with 100 subdivisions, is 0.333325.

2. The double sum has nine terms, each of the form $1/9 \sin(a) \sin(b)$, where (a, b) is the midpoint of one of the nine subrectangles. The total sum, evaluated numerically, is about 0.2133.

   Maple's value with $n = 10$, i.e., the midpoint sum with 100 subdivisions, is 0.211498.

3. The triple midpoint sum has eight terms, each of the form $8abc$, where (a, b, c) is the midpoint of one of the 8 subcubes. The total sum, evaluated numerically, is 512.

   Maple's value with $n = 2$, i.e., the triple midpoint sum with 8 subdivisions is 512. With $n = 4$, *Maple* still gives 512—this happens to be the exact answer.

4. (a) The level curves $z = 1$, $z = 2$, $z = 3$, etc., pass through the midpoints of the subdivisions. From this one concludes that the midpoint sum adds up to 64.

   (b) *Maple* agrees that the double midpoint sum is 64.

   (c) The level curves split each subrectangle in half; as a result, the midpoint sum commits no error in approximating the integral.

5. (a) Either the level curves or the formula can be used to estimate values of f at the midpoints; they lead to the estimate $I \approx M_{16} = 168$ (or something near that).

   (b) *Maple* calculates the double midpoint sum as 168.

   (c) Since the surface rises faster and faster as away from the origin, the midpoint sum somewhat underestimates the integral.

§M.6 Calculating Integrals by Iteration

1. If $R = [0, 1] \times [0, 1]$, then $\iint_R \sin(x) \sin(y) \, dA = \left(\cos(1)\right)^2 - 2 \cos(1) + 1 \approx 0.2113$.

2. If $R = [0, 1] \times [0, 1]$, then $\iint_R \sin(x + y) \, dA = -\sin(2) + 2 \sin(1) \approx 0.7736$.

3. If $R = [0, 4] \times [0, 4]$, then $\iint_R (x^2 + y^2) \, dA = \dfrac{512}{3} \approx 170.6667$.

4. If $V = [0, 1] \times [0, 2] \times [0, 3]$, then $\iiint_R x \, dV = 3$.

5. If $V = [0, 1] \times [0, 2] \times [0, 3]$, then $\iiint_R y \, dV = 6$.

6. $\iint_R (x + y) \, dA = \displaystyle\int_{x=0}^{x=1} \int_{y=x^2}^{y=x} (x + y) \, dy \, dx = \dfrac{3}{20} = 0.15$.

7. $\iint_R x \, dA = \displaystyle\int_{x=0}^{x=1} \int_{y=x^2}^{y=\sqrt{x}} x \, dy \, dx = \dfrac{3}{20} = 0.15$.

8. $\iint_R 1 \, dA = \displaystyle\int_{x=0}^{x=1} \int_{y=0}^{y=\sqrt{1-x^2}} 1 \, dy \, dx = \dfrac{\pi}{4} \approx 0.7854$.

9. $\iint_R (x + y) \, dA = \displaystyle\int_{y=0}^{y=1} \int_{x=y}^{x=\sqrt{y}} (x + y) \, dx \, dy = \dfrac{3}{20} = 0.15$.

10. $\iint_R x \, dA = \displaystyle\int_{y=0}^{y=1} \int_{x=y^2}^{x=\sqrt{y}} x \, dx \, dy = \dfrac{3}{20} = 0.15$.

11. $\iint_R 1 \, dA = \displaystyle\int_{y=0}^{y=1} \int_{x=0}^{x=\sqrt{1-y^2}} 1 \, dx \, dy = \dfrac{\pi}{4} \approx 0.7854$.

12. (a) Integrating first in y, then in x, $\iint_R (x + y) \, dA = \displaystyle\int_{x=-1}^{x=1} \int_{y=x^2}^{y=1} (x + y) \, dy \, dx = 4/5$.

 (b) Integrating first in x, then in y, $\iint_R (x + y) \, dA = \displaystyle\int_{y=0}^{y=1} \int_{x=-\sqrt{y}}^{x=\sqrt{y}} (x + y) \, dy \, dx = 4/5$.

13. (a) Integrating first in y, then in x, $\iint_R x \, dA = \displaystyle\int_{x=0}^{x=1} \int_{y=0}^{y=e^x} x \, dy \, dx = 1$.

 (b) Integrating first in x, then in y, requires that the y-interval be broken up into two parts:

$$\iint_R x \, dA = \int_{y=0}^{y=1} \int_{x=0}^{x=1} x \, dx \, dy + \int_{y=1}^{y=e} \int_{x=\ln y}^{x=1} x \, dx \, dy = 1/2 + 1/2 = 1.$$

14. (a) Single-variable calculus say that the area is the ordinary integral $\int_a^b f(x)\,dx$.

(b) Writing the double integral in iterated form gives

$$\iint_R 1\,dA = \int_{x=a}^{x=b} \int_{y=0}^{y=f(x)} 1\,dy\,dx = \int_{x=a}^{x=b} y\Big]_{y=0}^{y=f(x)} dx = \int_{x=a}^{x=b} f(x)\,dx;$$

this is the same formula as in one-variable calculus.

15. (a) Since R is the plane region bounded by the curves $x = g(y)$, $x = 0$, $y = c$, and $y = d$, the area of R is $\int_c^d g(y)\,dy$.

(b) $I = \iint_R 1\,dA = \int_c^d \int_0^{g(y)} 1\,dx\,dy = \int_c^d x\Big]_0^{g(y)} dy = \int_c^d g(y)\,dy.$

§M.7   Double Integrals in Polar Coordinates

1. (a) The area of a wedge with outer radius r_2 and making angle $\Delta\theta$ at the origin is $\dfrac{r_2{}^2}{2}\Delta\theta$.
        A polar rectangle is the difference of two such wedges; the result follows with a little computation.

    (b) If we set $a = r$, $b = r + \Delta r$, and $\beta - \alpha = \Delta\theta$ in part (a), then the result follows.

2. The answer is 2/3 in all cases. As iterated integrals:

    (a) $I = \displaystyle\int_{x=-1}^{x=1}\left(\int_{y=0}^{y=\sqrt{1-x^2}} y\,dy\right)dx.$

    (b) $I = \displaystyle\int_{y=0}^{y=1}\left(\int_{x=-\sqrt{1-y^2}}^{x=\sqrt{1-y^2}} y\,dx\right)dy.$

    (c) $I = \displaystyle\int_{\theta=0}^{\theta=\pi}\left(\int_{0}^{1} r^2\sin\theta\,dr\right)d\theta.$

3. (a) The solid lies under a surface with parabolic cross sections.

    (b) $I_1 = 2\pi/3$, no matter how you slice it.

4. The cardioid has area $3\pi/2$.

5. The triangle has area 1/2; note that it's bounded by the polar curves $\theta = 0$, $\theta = \pi/4$, and $r = \sec\theta$.

6. The circle in question has polar equation $r = \sin\theta$, for $0 \le \theta \le \pi$. The area inside is $\pi/4$.

7. The integral reduces to $\displaystyle\int_0^\pi (1 + \sin\theta)\,d\theta = \pi + 2.$

8. The surface lies above the unit circle in the xy-plane; the surface is defined by the function $g(r, \theta) = 1 - r^2$. Thus, in polar form, the integral is $\displaystyle\int_0^{2\pi}\int_0^1 (1 - r^2)r\,dr\,d\theta = \dfrac{\pi}{2}.$

9. The surface lies above the unit circle in the xy-plane; the surface is defined by the function $g(r, \theta) = 1 - r$. Thus, in polar form, the integral is $\displaystyle\int_0^{2\pi}\int_0^1 (1 - r)r\,dr\,d\theta = \dfrac{\pi}{3}.$